Introduction to Multiscale Mathematical Modeling

Introduction to Multiscale Mathematical Modeling

Grigory Panasenko
Vilnius University, Lithuania

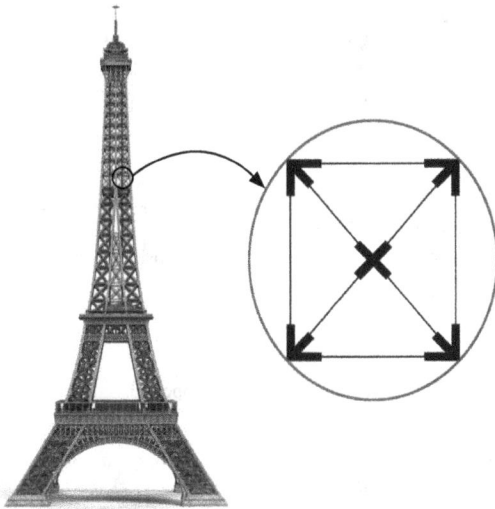

World Scientific

NEW JERSEY · LONDON · SINGAPORE · BEIJING · SHANGHAI · HONG KONG · TAIPEI · CHENNAI · TOKYO

Published by

World Scientific Publishing Europe Ltd.

57 Shelton Street, Covent Garden, London WC2H 9HE

Head office: 5 Toh Tuck Link, Singapore 596224

USA office: 27 Warren Street, Suite 401-402, Hackensack, NJ 07601

Library of Congress Cataloging-in-Publication Data
Names: Panasenko, Grigory, author.
Title: Introduction to multiscale mathematical modeling /
    Grigory Panasenko, Vilnius University, Lithuania.
Description: New Jersey : World Scientific, [2022] | Includes bibliographical references and index.
Identifiers: LCCN 2022011413 | ISBN 9781800612310 (hardcover) |
    ISBN 9781800612327 (ebook for institutions) | ISBN 9781800612334 (ebook for individuals)
Subjects: LCSH: Multiscale modeling. | Mathematical models.
Classification: LCC QA401 .P28 2022 | DDC 511/.8--dc23/eng20220422
LC record available at https://lccn.loc.gov/2022011413

British Library Cataloguing-in-Publication Data
A catalogue record for this book is available from the British Library.

For any available supplementary material, please visit
https://www.worldscientific.com/worldscibooks/10.1142/Q0363#t=suppl

Desk Editors: Balamurugan Rajendran/Adam Binnie/Shi Ying Koe

Typeset by Stallion Press
Email: enquiries@stallionpress.com

Preface

This book is based on the courses given for the master's and Ph.D. students of 2015–2021 batch at Skoltech (Moscow), University of Chile (Santiago), University Jean Monnet (Saint-Etienne, France), University of Benevento (Italy), and Politecnico di Torino. The third chapter was given as a mini-course at the 9th International Moscow Winter School in Physics. The book introduces main mathematical models describing mechanical behavior at microscopic level of heterogeneous media and for blood flow in a network of vessels. Homogenization technique is applied for multiscale analysis of heterogeneous media. For the network of vessels, asymptotic methods (matching, boundary layers) are presented. The method of asymptotic partial decomposition of the domain defines hybrid dimension models combining one-dimensional (1-D) description obtained from the dimension reduction with three-dimensional (3-D) zooms. It justifies the special exponentially precise junction conditions at the interface of 1-D and 3-D parts. It can be applied to model the blood flow in vessels with a thrombus or a stent. The course contains an important introductory part recalling the necessary mathematical background so that it is accessible not only for the master's and Ph.D. students in mathematics but also for those in engineering and biophysics.

The structure of the book is as follows. The first chapter introduces the main equations of mathematical physics. These equations are derived from the conservation laws. The main types of boundary conditions are introduced. The second chapter contains the minimal mathematical background which is required for better understanding of the multiscale methods. It recalls the main notions of analysis, in particular, of functional analysis. These notions are applied to prove the existence, uniqueness, and *a priori* estimates for the partial derivative equations derived in Chapter 1. These theorems are used to prove the error estimates for the multiscale approximations.

The third chapter introduces the homogenization technique as a general method used to pass from the microscopic scale to the macroscopic scale. First, this method is presented for a 1-D heat equation. Then, a similar presentation is developed in the case of the conductivity equation in multiple dimensions.

Then, the error estimates are derived for the approximations of the homogenization method. These estimates are crucial to persuade the users that the method is effective and to fix the limitations of the theory. It contains the general scheme of the proof of estimates justifying the asymptotic expansions.

Finally, Chapter 4 is devoted to the asymptotic analysis of thin domains and the method of asymptotic partial decomposition of domain. This method allows to combine and glue models of different dimensions to reduce the dimension in an important part of the domain keeping the 3-D zooms. The solution of such hybrid dimension model almost coincides with the exact solution but needs much less computational resources.

About the Author

Grigory Panasenko is a Distinguished Professor graduated from Moscow State University M.V. Lomonosov in 1976 where he was sequentially Ph.D. student, Associate Professor, and Professor. In 1993, he began working at the University Jean Monnet (Saint-Etienne, France). And since 2018, he has been serving in a Main Researcher position at Vilnius University. Professor Panasenko is an author of 3 books and 150 articles in applied mathematics.

Contents

Chapter 1

Derivation of the Main Equations of Mathematical Physics

1. Heat Equation

The derivation of the heat equation is based on the heat (energy) conservation law and Fourier's thermal conductivity law.

The *energy conservation law* states that in a small fixed volume V (Lagrangian volume), the change (accumulation) of the energy Q from the moment t_0 to the moment $t_0 + \tau$ is equal to the sum of the energy produced inside the volume and the input of the energy through the boundary of the volume. The change in the energy is proportional to the change in the temperature u:

$$\Delta Q = c\rho \Delta u, \qquad (1.1.1)$$

where c is the specific heat capacity and ρ is the mass density of the material.

Denote $f(x,t)$ as the energy output per unit time and unit volume on point $x = (x_1, x_2, x_3)$ in space and at moment t, and let $\mathcal{F}(x,t)$ ($\mathcal{F} = (\mathcal{F}_1, \mathcal{F}_2, \mathcal{F}_3)$) be the flux on point x at moment t. Here, $\mathcal{F}_i$ $i = 1,2,3$, is the flow of energy per unit of area per unit of time through a small surface perpendicular to the Cartesian axis Ox_i.

1

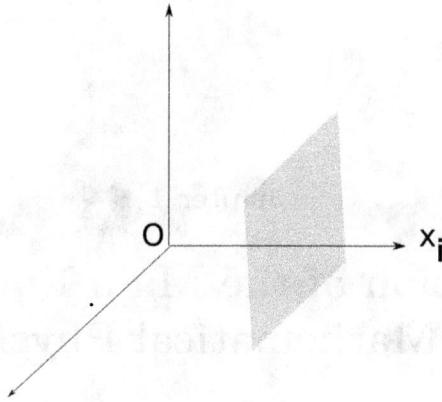

Consider the Lagrangian volume

$$V = C_{x_0,h} = \left(x_{01} - \frac{h}{2}, x_{01} + \frac{h}{2}\right) \times \left(x_{02} - \frac{h}{2}, x_{02} + \frac{h}{2}\right)$$

$$\times \left(x_{03} - \frac{h}{2}, x_{03} + \frac{h}{2}\right),$$

where $x_0 = (x_{01}, x_{02}, x_{03})$.

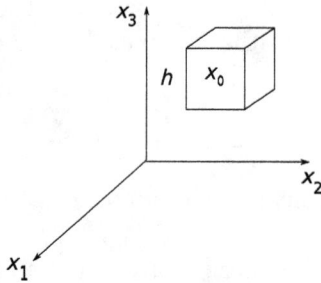

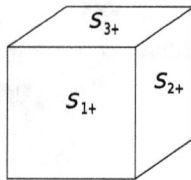

Denote squares as

$$S_{1\pm} = \left\{ x = (x_1, x_2, x_3) \mid x_1 = x_{10} \pm \frac{h}{2}, \right.$$

$$\left. x_2 \in \left(x_{02} - \frac{h}{2}, x_{02} + \frac{h}{2} \right), \quad x_3 \in \left(x_{03} - \frac{h}{2}, x_{03} + \frac{h}{2} \right) \right\},$$

$$S_{2\pm} = \left\{ x = (x_1, x_2, x_3) \mid x_2 = x_{20} \pm \frac{h}{2}, \right.$$

$$\left. x_1 \in \left(x_{01} - \frac{h}{2}, x_{01} + \frac{h}{2} \right), \quad x_3 \in \left(x_{03} - \frac{h}{2}, x_{03} + \frac{h}{2} \right) \right\},$$

$$S_{3\pm} = \left\{ x = (x_1, x_2, x_3) \mid x_3 = x_{30} \pm \frac{h}{2}, \right.$$

$$\left. x_1 \in \left(x_{01} - \frac{h}{2}, x_{01} + \frac{h}{2} \right), \quad x_2 \in \left(x_{02} - \frac{h}{2}, x_{02} + \frac{h}{2} \right) \right\}.$$

Let $\mathbf{n}$ be the outer normal of V. Then, (1.1.1) and the conservation law take the form

$$\int_V c(x)\rho(x)(u(x, t_0 + \tau) - u(x, t_0))dx$$

$$= \int_{t_0}^{t_0+\tau} \int_V f(x, t)dxdt - \sum_{i=1}^{3} \sum_{\pm} \int_{t_0}^{t_0+\tau} \int_{S_{i\pm}} \mathcal{F}(x, t) \cdot \mathbf{n}dsdt. \quad (1.1.2)$$

Divide (1.1.2) by τh^3. Assume that all functions in the integral terms of (1.1.2) are continuous and apply the mean value theorem: if Π is a closed parallelepiped in $\mathbb{R}^d$ and F is a continuous function $\Pi \to \mathbb{R}$, then there exists a point $y_* \in \Pi$ such that

$$\int_\Pi F(y)dy = F(y_*)\text{mes}\Pi.$$

We get

$$c(x_*)\rho(x_*)\frac{u(x_*, t_0 + \tau) - u(x_*, t_0)}{\tau} = f(x_{**}, t_{**})$$

$$-\frac{\mathcal{F}_1\left(x_{01} + \frac{h}{2}, x_{2*}^{(1)}, x_{3*}^{(1)}\right) - \mathcal{F}_1\left(x_{01} - \frac{h}{2}, x_{2*}^{(1)}, x_{3*}^{(1)}\right)}{h}$$

$$-\frac{\mathcal{F}_2\left(x_{1*}^{(2)}, x_{02} + \frac{h}{2}, x_{3*}^{(2)}, t_*^{(2)}\right) - \mathcal{F}_2\left(x_{1*}^{(2)}, x_{02} - \frac{h}{2}, x_{3*}^{(2)}, t_*^{(2)}\right)}{h}$$

$$-\frac{\mathcal{F}_3\left(x_{1*}^{(3)}, x_{2*}^{(3)}, x_{03} + \frac{h}{2}, t_*^{(3)}\right) - \mathcal{F}_3\left(x_{1*}^{(3)}, x_{2*}^{(3)}, x_{03} - \frac{h}{2}, t_*^{(3)}\right)}{h},$$

$$(1.1.3)$$

where x_*, $x_{**} \in V$, $x_{1*}^{(2)}, x_{1*}^{(3)} \in \left(x_{01} - \frac{h}{2}, x_{01} + \frac{h}{2}\right)$,
$x_{2*}^{(1)}, x_{2*}^{(3)} \in \left(x_{02} - \frac{h}{2}, x_{02} + \frac{h}{2}\right)$, $x_{3*}^{(1)}$, and $x_{3*}^{(2)} \in \left(x_{03} - \frac{h}{2}, x_{03} + \frac{h}{2}\right)$.

Let φ be differentiable in $[a, b]$, $y_0 \in (a, b)$, then

$$\lim_{h \to 0^+} \frac{\varphi\left(y_0 + \frac{h}{2}\right) - \varphi\left(y_0 - \frac{h}{2}\right)}{h}$$

$$= \lim_{h \to 0^+} \left\{\frac{\varphi\left(y_0 + \frac{h}{2}\right) - \varphi(y_0)}{h} + \frac{\varphi(y_0) - \varphi\left(y_0 - \frac{h}{2}\right)}{h}\right\}$$

$$= \frac{1}{2}\left\{\lim_{h \to 0^+} \frac{\varphi\left(y_0 + \frac{h}{2}\right) - \varphi(y_0)}{h/2} + \lim_{h \to 0^+} \frac{\varphi(y_0) - \varphi\left(y_0 - \frac{h}{2}\right)}{h/2}\right\}$$

$$= \frac{1}{2}\left\{\frac{d\varphi}{dy}(y_0) + \frac{d\varphi}{dy}(y_0)\right\} = \frac{d\varphi}{dy}(y_0).$$

So, passing to the limit as $h \to 0^+$, $\tau \to 0^+$ in (1.1.3), we get

$$c(x_0)\rho(x_0)\frac{\partial u}{\partial t}(x_0, t_0) = f(x_0, t_0) - \sum_{i=1}^{3} \frac{\partial \mathcal{F}_i}{\partial x_i}(x_0, t_0), \qquad (1.1.4)$$

i.e.

$$c(x)\rho(x)\frac{\partial u}{\partial t}(x, t) = f(x, t) - \mathrm{div}\mathcal{F}(x, t). \qquad (1.1.5)$$

The *Fourier's law* of thermal conductivity states

$$\mathcal{F} = -k(x)\nabla u(x,t), \tag{1.1.6}$$

where $k(x)$ is the *conductivity matrix*,

$$\nabla u = \begin{pmatrix} \frac{\partial u}{\partial x_1} \\ \frac{\partial u}{\partial x_2} \\ \frac{\partial u}{\partial x_3} \end{pmatrix}, \quad k = \begin{pmatrix} k_{11} & k_{12} & k_{13} \\ k_{21} & k_{22} & k_{23} \\ k_{31} & k_{32} & k_{33} \end{pmatrix},$$

which is symmetric ($k_{ij} = k_{ji}$) and positive definite:

$$\sum_{i,j=1}^{3} k_{ij}(x)\xi_j\xi_i \geq \kappa \sum_{i=1}^{3} \xi_i^2 \tag{1.1.7}$$

for all $(\xi_1, \xi_2, \xi_3) \in \mathbb{R}^3$ with $\kappa > 0$.

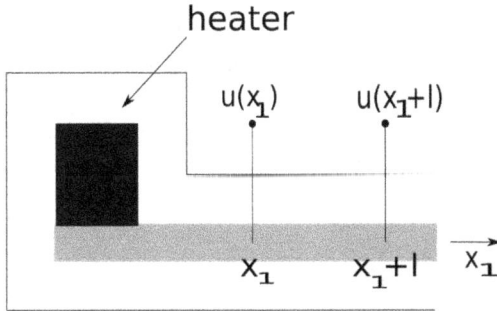

In Fourier's experiment, $\mathcal{F}_1 = k\frac{u(x_1)-u(x_1+l)}{l}$.

In an isotropic case, k is a scalar value.

Plug (1.1.6) in (1.1.5). We get the *heat equation*:

$$C(x)\frac{\partial u}{\partial t}(x,t) = \operatorname{div}(k(x)\nabla u) + f(x,t), \tag{1.1.8}$$

where $C(x) = c(x)\rho(x)$.

Here, u is the unknown function, and C, k, f are known data with usually prescribed regularity properties, for instance, differentiable (C^1) functions.

In anisotropic case, (1.1.8) has the form

$$C(x)\frac{\partial u}{\partial t}(x,t) = \sum_{i,j=1}^{3} \frac{\partial}{\partial x_i}\left(k_{ij}(x)\frac{\partial u}{\partial x_j}\right) - f(x,t). \qquad (1.1.9)$$

2.   Boundary/Initial/Interface Conditions

A *mathematical model* of some physical process is not only its corresponding partial differential equation (PDE), such as the heat equation (1.1.8), but it includes as well several boundary/initial/interface conditions ensuring the uniqueness of a solution. In particular, the heat equation (1.1.8) needs one (as many as the order of the time derivative) initial condition:

$$u\mid_{t=0}= \varphi(x), \qquad (1.2.1)$$

where φ is a given function.

Usually, the heat equation is considered in a bounded volume. Then, equation (1.1.8) needs one boundary condition on the whole *boundary ∂G* of the *domain G*.

Let us introduce some notions which will be used later.

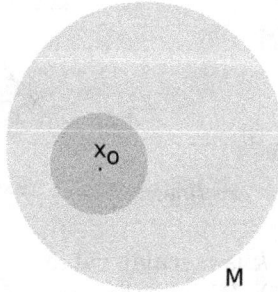

A set M of $\mathbb{R}^d$ $(d = 1, 2, 3)$ is called an *open set* if for any point $x_0 \in M$, there exists a ball[a] $B(x_0, r)$ belonging to M $(B(x_0, r) \subset M)$. A set M of $\mathbb{R}^d$ is called a *closed set* if for any sequence $(x_n)_{n \in \mathbb{N}}$,

---

[a] $B(x_0, r) = \{x \in \mathbb{R}^d \mid \|x - x_0\| < r\}$, where x_0 is the center, $r > 0$ is the radius.

$x_n \in M$, converging to some point $\bar{x}$, the limit point $\bar{x}$ belongs to M, i.e. $(x_n \to \bar{x}$ and $x_n \in M) \Rightarrow \bar{x} \in M$. A set M of $\mathbb{R}^d$ is called a *connected set* if any two points x, y of M can be connected by a continuous curve C_{xy} belonging to M $(C_{xy} \subset M)$.

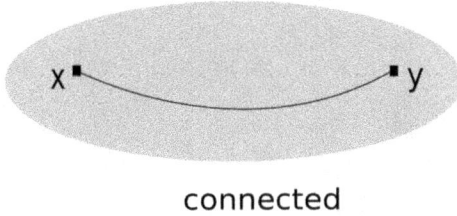

connected

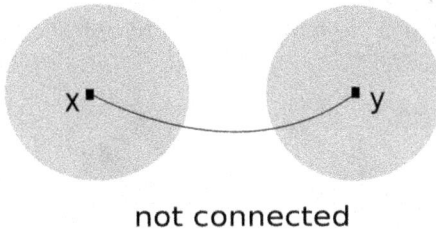

not connected

A set M of $\mathbb{R}^d$ is called a *bounded set* if there exists a ball $B(0, r)$ containing the whole set M $(M \subset B(0, r))$.

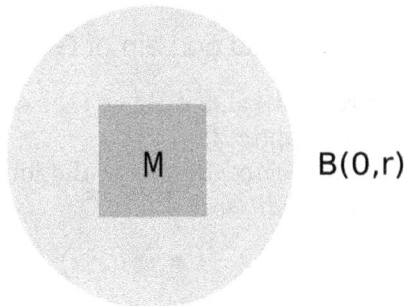

A set ∂M of $\mathbb{R}^d$ is called the *boundary* of a set $M \subset \mathbb{R}^d$ if any ball with the center at the points of ∂M contains at least one point belonging to M and at least one point out of M.

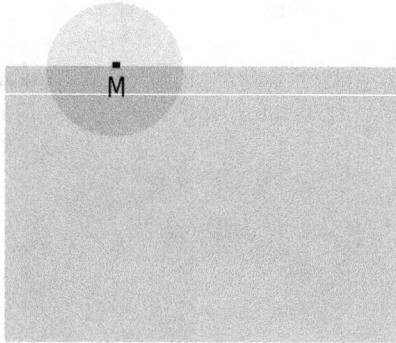

boundary

A set $\bar{M} = M \cup \partial M$ is called the *closure* of M. An open-connected set in $\mathbb{R}^d$ is called a *domain*. In the present section, we consider a bounded domain G.

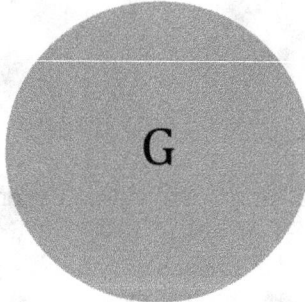

bounded domain of G

These definitions remain valid if $\mathbb{R}^d$ is replaced by any normed space (see Section 1 of Chapter 2).

We consider the *boundary conditions* of the following types.

Dirichlet's boundary condition:

$$u(x,t) \big|_{x \in \partial G} = \psi_D(x,t) \tag{1.2.2$_D$}$$

(given temperature at the boundary), or

Neumann's boundary condition:

$$-k(x)\nabla u(x,t) \cdot \mathbf{n} \big|_{x \in \partial G} = \psi_N(x,t) \tag{1.2.2$_N$}$$

(given normal flux at the boundary), where $\mathbf{n}$ is an outer normal vector and $\cdot$ denotes the Euclidean inner product,

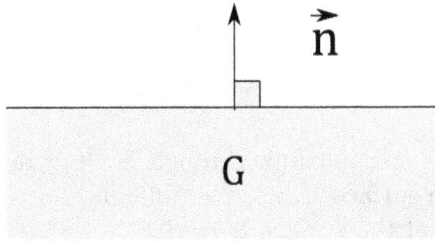

or

Robin's boundary condition:

$$-k(x)\nabla u(x,t) \cdot \mathbf{n} \mid_{x \in \partial G} = \lambda(x)u(x,t) \mid_{x \in \partial G} + \psi_R(x,t) \quad (1.2.2_R)$$

(the normal flux at the boundary is proportional to the jump of the temperature from the surface to the surrounding area), or **periodicity** condition if G is a cube, $G = (0,1)^d$:

u is a 1-periodic function defined on $\mathbb{R}^d$, i.e.

$$\forall x \in \mathbb{R}^d, \quad \forall z \in \mathbb{Z}^d,$$
$$u(x+z,t) = u(x,t). \quad (1.2.2_p)$$

Functions ψ_D, ψ_N, ψ_R, and λ are given. Also, the whole boundary ∂G can be subdivided into three parts Γ_D, Γ_N, and Γ_R such that $\Gamma_D \cup \Gamma_N \cup \Gamma_R = \partial G$ and $\Gamma_D \cap \Gamma_N = \emptyset$, $\Gamma_D \cap \Gamma_R = \emptyset$, $\Gamma_N \cap \Gamma_R = \emptyset$, and one sets boundary condition $(1.2.2_D)$ on Γ_D, boundary condition $(1.2.2_N)$ on Γ_N, and boundary condition $(1.2.2_R)$ on Γ_R.

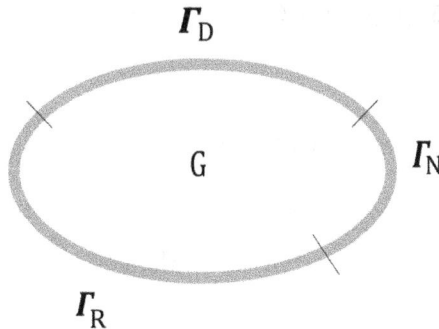

If k is discontinuous at some surface Σ, then at this surface, equation (1.1.8) is replaced by two interface conditions:

$$[u]_\Sigma = 0 \quad (1.2.3_1)$$

(continuity of the temperature) and

$$[-k(x)\nabla u \cdot \mathbf{n}]_\Sigma = 0 \qquad (1.2.3_2)$$

(continuity of the normal flux through Σ. $[u]_\Sigma$ denotes a jump of function u at the surface Σ, i.e. the difference of the limit values of u from two sides of the surface Σ).

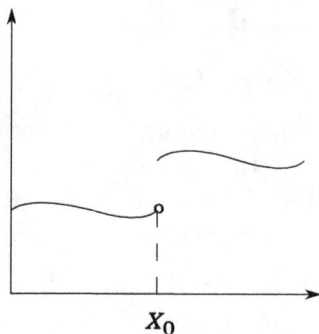

$$[u]_{x_0} = \lim_{x \to x_0+0} u(x) - \lim_{x \to x_0-0} u(x).$$

The heat propagation process is so described by a mathematical model consisting of the heat equation (1.1.8), initial condition (1.2.1), boundary condition (1.2.2), and eventually interface conditions (1.2.3).

3.   Particular Cases: Generalizations

3.1.   *Stationary equation*

If a process does not depend on time, then $\frac{\partial u}{\partial t} = 0$ and we obtain the *stationary conductivity equation*

$$-\mathrm{div}(k(x)\nabla u) = f(x). \qquad (1.3.1)$$

The stationary heat model consists of equation (1.3.1) and one of boundary conditions (1.2.2). Clearly, it doesn't need any initial condition.

3.2. *Two-dimensional and one-dimensional models*

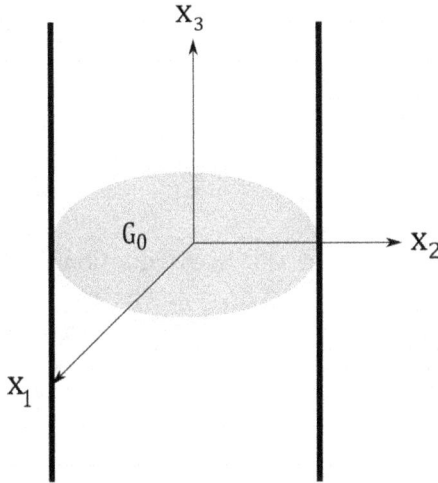

If a domain G is a cylinder $G_0 \times \mathbb{R}$, where G_0 is a bounded domain in $\mathbb{R}^2$, and if the right-hand side f in (1.1.8), coefficients C and k, and functions φ and ψ do not depend on x_3, then we can consider a solution $u(x,t)$ which is also independent of x_3. Equation (1.1.9) takes the form

$$C(x)\frac{\partial u}{\partial t}(x',t) = \sum_{i,j=1}^{2} \frac{\partial}{\partial x_i}\left(k_{ij}(x')\frac{\partial u}{\partial x_j}\right) + f(x',t), \qquad (1.3.2)$$

where $x' = (x_1, x_2)$.

If the domain G is a layer $(a,b) \times \mathbb{R}^2$ and the right-hand side f and the coefficients, functions φ and ψ do not depend on (x_2, x_3), then we get a one-dimensional model in the interval (a,b):

$$C(x_1)\frac{\partial u}{\partial t}(x_1,t) = \frac{\partial}{\partial x_1}\left(k_{11}(x_1)\frac{\partial u}{\partial x_1}\right) + f(x_1,t), \qquad (1.3.3)$$

where the subscript 1 can be omitted.

Another source of 2-D and 1-D models is the method of asymptotic dimension reduction for thin domains.

Finally, consider some generalizations of equation (1.1.8). It can have several additional terms:

$$C(x)\frac{\partial u}{\partial t} = \text{div}(k(x)\nabla u) - C(x)\mathbf{V}(x,t) \cdot \nabla u - \alpha(x)u + f(x,t),$$

(1.3.4)

where the second term in the right-hand side corresponds to a convection with velocity $\mathbf{V}$ and the third term to a sorption process. For a compressible fluid, the second term on the right-hand side of (1.3.4) is $C\text{div}(\mathbf{V}u)$.

Another modification of the heat equation (1.1.8) is related to a correction of the Fourier's law, namely, the conductivity k may depend on the temperature, so the relation between the flux and the temperature becomes nonlinear:

$$\mathcal{F} = -k(x,u)\nabla u(x,t).$$

Equation (1.1.8) as well as its generalizations can describe other physical processes, in particular, diffusion of a substance. Then, u is the concentration, $C = 1$, and k is the diffusion coefficient. This is why equation (1.1.8) is also called diffusion equation, while (1.3.4) is called diffusion–convection equation. Its derivation follows the same scheme as was applied in Section 1. The energy conservation law is replaced by the mass conservation law, and the law (1.1.6) is called Fick's law. The stationary form (1.3.1) is used for the pressure field in the porous medium, where u is the pressure, k is the permeability coefficient, and Fourier's law becomes Darcy's law ($\mathcal{F}$ is the average velocity field).

This section shows that the same mathematical equation can describe physical processes of different nature.

Finally, consider the case of an isotropic homogeneous material having a constant conductivity k. Then, the term $\text{div}k(x)\nabla u$ becomes $k\Delta u$, where Δ is the Laplace operator (*Laplacian*):

$$\Delta = \sum_{i=1}^{d} \frac{\partial^2}{\partial x_i^2}.$$

Then, equation (1.1.8) reads

$$C\frac{\partial u}{\partial t} = k\Delta u + f,$$

and the normal flux $-k\nabla u \cdot \mathbf{n}$ reads $-k\frac{\partial u}{\partial n}$.

4. Elasticity Equation (Solid Mechanics)

Let us introduce the main notions of elasticity theory, considering deformation as a transformation of an elastic body from a reference configuration to a current configuration.

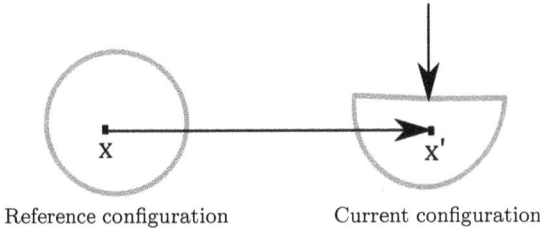

Reference configuration   Current configuration

Let a material point P of the reference configuration having coordinates $x = (x_1, x_2, x_3)$ move to the point $P'(x')$, $x' = (x_1', x_2', x_3')$ of the current configuration. The *displacement* vector $\mathbf{u}(x, t)$ is defined as the vector $\mathbf{PP'}$, connecting points $\mathbf{P}$ and $\mathbf{P'}$.

Strain tensor describes deformation in terms of relative displacement, i.e.

$$e = (e_{lj}(\mathbf{u}))_{1 \le l, j \le 3},$$

$$e_{lj}(\mathbf{u}) = \frac{1}{2} \left(\frac{\partial u_l}{\partial x_j} + \frac{\partial u_j}{\partial x_l} + \sum_{m=1}^{3} \frac{\partial u_m}{\partial x_j} \frac{\partial u_m}{\partial x_l} \right). \qquad (1.4.1)$$

However, the linear elasticity considers a **"small strain"** relation, neglecting the last term in (1.4.1):

$$e_{lj}(\mathbf{u}) = \frac{1}{2} \left(\frac{\partial u_l}{\partial x_j} + \frac{\partial u_j}{\partial x_l} \right). \qquad (1.4.1_{\mathrm{L}})$$

Define the *stress tensor* in the point $x_0 = (x_{01}, x_{02}, x_{03})$:

$$\sigma = (\sigma_{ik})_{1 \le i, r \le 3}.$$

This matrix consists of three columns: $\boldsymbol{\sigma}^{(1)}$, $\boldsymbol{\sigma}^{(2)}$, and $\boldsymbol{\sigma}^{(3)}$. In order to define these columns, consider an imaginary surface S_k, $k = 1, 2, 3$ passing through the point x_0.

Let S_1 be square

$$\left\{ (x_1, x_2, x_3) \mid x_1 = x_{01}, (x_2, x_3) \in \left(x_{02} - \frac{h}{2}, x_{02} + \frac{h}{2} \right) \right.$$
$$\left. \times \left(x_{03} - \frac{h}{2}, x_{03} + \frac{h}{2} \right) \right\}$$

perpendicular to the axis Ox_1.

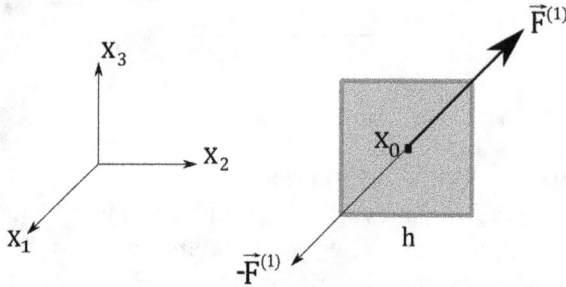

Assume that there is an imaginary incision at S_1 surface. Surface S_1 divides the continuous body into two segments. Consider the force $\mathbf{F}^{(1)}$ applied to one side of S_1 which is the surface of the segment $x_1 \leq x_{01}$. According to Newton's third law, the force applied to the second side of the surface S_1 is $-\mathbf{F}^{(1)}$.

The column $\boldsymbol{\sigma}^{(1)}$ is defined as follows:

$$\boldsymbol{\sigma}^{(1)} = \lim_{h \to 0} \frac{\mathbf{F}^{(1)}}{h^2}.$$

In a similar way, we consider the square incisions

$$S_2 = \left\{ (x_1, x_2, x_3) \mid x_1 \in \left(x_{01} - \frac{h}{2}, x_{01} + \frac{h}{2} \right), \right.$$
$$\left. x_2 = x_{02}, x_3 \in \left(x_{03} - \frac{h}{2}, x_{03} + \frac{h}{2} \right) \right\},$$

$$S_3 = \left\{ (x_1, x_2, x_3) \mid (x_1, x_2) \in \left(x_{01} - \frac{h}{2}, x_{01} + \frac{h}{2} \right) \right.$$
$$\left. \times \left(x_{02} - \frac{h}{2}, x_{02} + \frac{h}{2} \right), \quad x_3 = x_{03} \right\},$$

and define $\boldsymbol{\sigma}^{(2)}$, $\boldsymbol{\sigma}^{(3)}$.

The strain tensor e and the stress tensor σ are symmetric matrices.

The derivation of the linear elasticity equation is based on the consideration of a small cube $V = C_{x_0,h}$ from Section 1. Applying Newton's second law, we get for the acceleration of the cube multiplied by density ρ the following integral relation:

$$\int_V \rho(x)\frac{\partial^2 \mathbf{u}}{\partial t^2}dx = \int_V \mathbf{f}(x,t)dx$$

$$+ \sum_{i=1}^{3}\left\{\int_{S_{i+}} \boldsymbol{\sigma}^{(i)}(x,t)ds - \int_{S_{i-}} \boldsymbol{\sigma}^{(i)}(x,t)ds\right\}, \quad (1.4.2)$$

where $\mathbf{f}(x,t)$ is a given mass force "diffused" in the volume.

As in Section 1, we make a passage to the limit in (1.4.2) as $h \to 0$ and get the equation

$$\rho(x)\frac{\partial^2 \mathbf{u}}{\partial t^2} = \operatorname{div}\sigma + \mathbf{f}(x,t), \quad (1.4.3)$$

where

$$\operatorname{div}\sigma = \sum_{i=1}^{3}\frac{\partial}{\partial x_i}\boldsymbol{\sigma}^{(i)}.$$

The stress σ is related to the strain e by Hooke's law (constitutive equation):

$$\sigma_{i,k} = \sum_{j,l=1}^{3} a_{ij}^{kl}(x)e_{jl}(\mathbf{u}), \quad (1.4.4)$$

where a_{ij}^{kl} is the stiffness tensor (elasticity moduli).

Replacing $e_{jl}(\mathbf{u})$ in (1.4.4) by the expressions $(1.4.1)_{\mathrm{L}}$ and plugging (1.4.4) into (1.4.3), we get the *elasticity equation*

$$\rho(x)\frac{\partial^2 \mathbf{u}}{\partial t^2} = \sum_{j,l=1}^{3}\frac{\partial}{\partial x_i}\left(A_{ij}(x)\frac{\partial \mathbf{u}}{\partial x_j}\right) + \mathbf{f}(x,t), \quad (1.4.5)$$

where A_{ij} are 3×3 matrices with entries $\left(a_{ij}^{kl}\right)_{1\leq k,l\leq 3}$.

An isotropic material is characterized by two constants: *Young's modulus* E and *Poisson's ratio* ν. Then, a_{ij}^{kl} are defined as follows:

$$a_{ij}^{kl} = \frac{E}{2(1+\nu)} \left(\frac{2\nu}{1-2\nu} \delta_{ik}\delta_{jl} + \delta_{ij}\delta_{kl} + \delta_{il}\delta_{jk} \right), \qquad (1.4.6)$$

where $\delta_{ij} = \begin{cases} 1 \text{ if } i=j \\ 0 \text{ if not} \end{cases}$ is Kronecker's delta, $E > 0$, and $-1 < \nu < \frac{1}{2}$.

An isotropic material is also characterized by Lame's constants, λ and μ (μ is called the shear modulus). These constants are related to E and ν:

$$\lambda = \frac{E\nu}{(1+\nu)(1-2\nu)}, \quad \mu = \frac{E}{2(1+\nu)}.$$

In the anisotropic (orthotropic) case, the material is characterized by nine constants: three Young's moduli E_1, E_2, E_3; three Poisson's ratios ν_{21}, ν_{31}, ν_{32}; three shear moduli μ_{21}, μ_{31}, μ_{32}. Then, a_{ij}^{kl} are defined by the following relations:

$$\begin{pmatrix} a_{11}^{11} & a_{12}^{12} & a_{13}^{13} \\ a_{12}^{12} & a_{22}^{22} & a_{23}^{23} \\ a_{13}^{13} & a_{23}^{23} & a_{33}^{33} \end{pmatrix} = \begin{pmatrix} 1/E_1 & -\nu_{21}/E_2 & -\nu_{31}/E_3 \\ -\nu_{21}/E_2 & 1/E_2 & -\nu_{32}/E_3 \\ -\nu_{31}/E_3 & -\nu_{32}/E_3 & 1/E_3 \end{pmatrix}^{-1},$$

$$a_{11}^{22} = \mu_{21}, \quad a_{11}^{33} = \mu_{31}, \quad a_{22}^{33} = \mu_{32}, \qquad (1.4.7)$$

where $a_{ij}^{kl} = 0$ if one of the indices i, j, k, l is different from all others: $a_{12}^{22} = 0$. Also,

$$a_{ij}^{kl} = a_{kj}^{il} = a_{ji}^{lk} = a_{il}^{kj}. \qquad (1.4.8)$$

Equation (1.4.5) needs two initial conditions:

$$\mathbf{u}\,|_{t=0} = \boldsymbol{\varphi}_0(x), \qquad (1.4.9_0)$$

$$\frac{\partial \mathbf{u}}{\partial t}\,|_{t=0} = \boldsymbol{\varphi}_1(x), \qquad (1.4.9_1)$$

where $\boldsymbol{\varphi}_0$ and $\boldsymbol{\varphi}_1$ are given initial displacement and initial velocity, respectively.

The boundary condition is similar to (1.2.2):

$$\mathbf{u} \,|_{\partial G} = \boldsymbol{\psi}_D(x, t) \qquad (1.4.10_D)$$

(given displacement)
   or (if $\mathbf{n} = (n_1, n_2, n_3)$ is an outer normal vector)

$$\sum_{i,j=1}^{3} n_i A_{ij} \frac{\partial \mathbf{u}}{\partial x_j} \,|_{\partial G} = \boldsymbol{\psi}_N(x, t) \qquad (1.4.10_N)$$

(given normal stress)

or

$$\sum_{i,j=1}^{3} n_i A_{ij} \frac{\partial \mathbf{u}}{\partial x_j} \,|_{\partial G} = -\lambda(x)\mathbf{u} \,|_{\partial G} + \boldsymbol{\psi}_R(x, t) \qquad (1.4.10_R)$$

(springs at the boundary)
   or the periodicity condition.
   In the case of discontinuous A_{ij}, we have at the surface of discontinuity

$$[\mathbf{u}]_\Sigma = 0,$$

$$\left[\sum_{i,j=1}^{3} n_i A_{ij} \frac{\partial \mathbf{u}}{\partial x_j} \right]_\Sigma = 0. \qquad (1.4.11)$$

Here, $\mathbf{n} = (n_1, n_2, n_3)$ is the normal vector.
   As in Section 3, we can consider the *stationary elasticity equation*

$$-\sum_{i,j=1}^{3} \frac{\partial}{\partial x_i} \left(A_{ij}(x) \frac{\partial \mathbf{u}}{\partial x_j} \right) = \mathbf{f}(x), \quad x \in G, \qquad (1.4.12)$$

supplied with the boundary condition (1.4.10) or periodicity condition.

Also, we can consider the 2-D or 1-D version. Finally, the nonlinear elasticity equation (stationary case) is

$$-\sum_{i=1}^{d} \frac{\partial}{\partial x_i} \mathbf{A}_i(\nabla \mathbf{u}, x) = \mathbf{f}(x), \qquad (1.4.13)$$

where $\mathbf{A}_i$ are given vector-valued functions. Its form can correspond to the modifications of the scheme of the derivation of equation (1.4.5). Namely, we can replace relation (1.4.1$_L$) by (1.4.1). Then, leaving the remaining part of the derivation without any change, we get the so-called geometrically nonlinear elasticity equation:

$$-\frac{1}{2}\sum_{i,j,l=1}^{3}\frac{\partial}{\partial x_i}\left\{a_{ij}^{kl}(x)\left(\frac{\partial u_l}{\partial x_j}+\frac{\partial u_j}{\partial x_l}+\sum_{m=1}^{3}\frac{\partial u_m}{\partial x_j}\frac{\partial u_m}{\partial x_l}\right)\right\}=f_k(x),$$

$$k=1,2,3. \qquad (1.4.14)$$

Another possibility of modification is nonlinear Hooke's law (in particular, hyperelastic materials). The most popular model of a **physically** nonlinear isotropic material is as follows. Define the traces of e and σ as

$$e(1)=\sum_{i=1}^{3}e_{ii}, \quad \sigma(1)=\sum_{i=1}^{3}\sigma_{ii},$$

the deviators of strains and stresses as

$$E_{ij}=e_{ij}-\delta_{ij}(e(1)/3),$$

$$S_{ij}=\sigma_{ij}-\delta_{ij}(\sigma(1)/3), \qquad (1.4.15)$$

and the norms of deviators as

$$\hat{E}=\sqrt{\frac{1}{2}\sum_{i,j=1}^{3}E_{ij}^2},$$

$$\hat{S}=\sqrt{\frac{1}{2}\sum_{i,j=1}^{3}S_{ij}^2}. \qquad (1.4.16)$$

The nonlinear Hooke's law states the relation

$$\hat{S}=\varphi(\hat{E})\hat{E}, \qquad (1.4.17)$$

where φ is a given function, $\varphi(0)=2\mu$, where μ is the shear modulus,

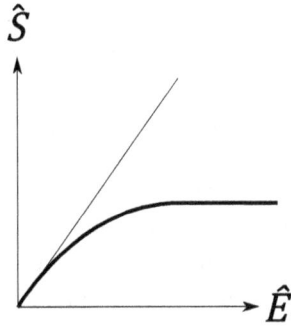

and

$$S_{ij} = \varphi(\hat{E})E_{ij}, \quad \sigma(1) = Ke(1), \qquad (1.4.18)$$

where

$$K = \lambda + \frac{2}{3}\mu = \frac{E}{3(1-2\nu)}$$

is the bulk modulus.

This physical nonlinearity can be combined with the geometrical nonlinearity.

The elasticity equation can be generalized if the Hooke's law states the linear proportionality of the stress tensor σ to the strain and to the strain rate tensors e and $\dot{e}$ ($\dot{e}$ denotes $\frac{\partial}{\partial t}e$), namely,

$$\sigma_{ik} = \sum_{j,l=1}^{3}\left(a_{ij}^{kl}(x)e_{jl}(\mathbf{u}) + b_{ij}^{kl}(x)\dot{e}_{jl}(\mathbf{u})\right) \qquad (1.4.19)$$

instead of (1.4.4).

It gives the visco-elasticity equation with short memory:

$$\rho(x)\frac{\partial^2\mathbf{u}}{\partial t^2} = \sum_{i,j=1}^{3}\frac{\partial}{\partial x_i}\left(A_{ij}(x)\frac{\partial\mathbf{u}}{\partial x_j}\right)$$

$$+ \sum_{i,j=1}^{3}\frac{\partial}{\partial x_i}\left(B_{ij}(x)\frac{\partial^2\mathbf{u}}{\partial t\partial x_j}\right) + f(x,t), \qquad (1.4.20)$$

where B_{ij} are 3×3 given matrices.

All elasticity models suppose a full reversibility of the change of shape in response to applied forces. If the deformations are irreversible, then the process is called plastic deformation. In this case, the stress–strain diagram contains a hysteresis loop. Mostly, metals respect the linear elastic behavior for tensile loading for strain e_{11} less than 0.002, after which the behavior becomes nonlinear elastic behavior up to the elastic limit. For strain greater than the elastic limit, the behavior becomes plastic; however, if loading continues it corresponds to the nonlinear elastic model.

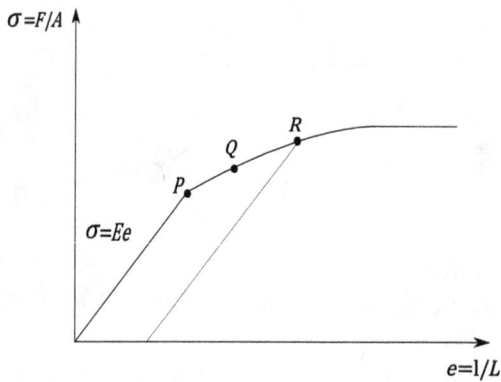

Stress–strain curve showing typical yield behavior: P is the limit of the linear Hooke's law, Q the elastic limit, and R the offset yield strength.

The following picture presents the Hooke's law experiment.

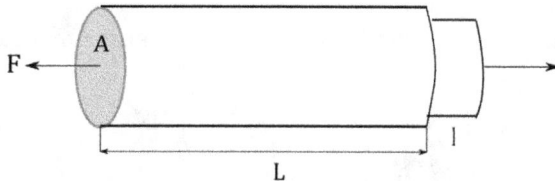

l is the length of extension,
L is the initial length of the bar,
A is the area of the cross-section, and
F is the applied tensile force.

5. Navier–Stokes and Stokes Equations (Fluid Mechanics)

Changes in properties of a moving medium can be measured in two different ways. One can measure a given property by either carrying out the measurement on a fixed point (or volume) as particles pass by or by following a parcel along a streamline.

The first approach is called Eulerian while the second is called Lagrangian. In the Lagrangian approach, the values are functions of the coordinates $x = (x_1, x_2, x_3)$ of a point at the initial moment of time (for example, before deformation, as in Section 4). In the Eulerian approach, all values are functions of the coordinates $x = (x_1, x_2, x_3)$ of the point where the observer stands. This point is fixed in space while the medium passes through. In this second approach, if the medium passes through a point $P(x_1, x_2, x_3)$ (point P having coordinates (x_1, x_2, x_3)) with the velocity $\mathbf{v}(x, t)$ and we consider a parcel V, then in time interval Δt, any value A within V changes from $A(x, t)$ to $A(x + \mathbf{v}(x, t)\Delta t, t + \Delta t)$ because the point $P(x)$ moves to $P'(x + \mathbf{v}(x, t)\Delta t)$. So, the velocity of the change of A is given by

$$\lim_{\Delta t \to 0} \frac{A(x + \mathbf{v}(x, t)\Delta t, t + \Delta t) - A(x, t)}{\Delta t}$$

$$= \frac{\partial A}{\partial t}(x, t) + \sum_{i=1}^{3} v_i(x, t) \frac{\partial A}{\partial x_i}(x, t) = \left(\frac{\partial}{\partial t} + \mathbf{v} \cdot \nabla \right) A.$$

This derivative is called material derivative denoted as DA/Dt.

Consider the motion of an incompressible fluid having density ρ. Then, applying the derivation scheme from Section 4 but in the frame of the Eulerian approach, we get the equation for Newton's second law:

$$\rho \frac{D\mathbf{v}}{Dt} = \operatorname{div} \sigma + \mathbf{f}(x, t), \tag{1.5.1}$$

which replaces (1.4.3).

The constitutive equation for a fluid is

$$\sigma = \mu D(\mathbf{v}) - pI, \tag{1.5.2}$$

where $D(\mathbf{v}) = \nabla \mathbf{v} + (\nabla \mathbf{v})^T$ is the symmetrized gradient of the velocity, μ is the dynamic viscosity, p is the pressure, I is the identity matrix, $\mathbf{f}$ is the distribution of a mass force (often $\mathbf{f} = \rho\mathbf{g}$, where $\mathbf{g}$ is the gravity acceleration). Note that $\nu = \mu/\rho$ is called kinematic viscosity.

The incompressibility of the fluid is described by the divergence equation:

$$\mathrm{div}\mathbf{v} = 0. \tag{1.5.3}$$

It follows from the mass continuity equation.

Plug (1.5.2) into (1.5.1). We get the Navier–Stokes equation:

$$\begin{cases} \rho\frac{\partial \mathbf{v}}{\partial t} = \mathrm{div}(\mu D(\mathbf{v}) - pI) - \rho(\mathbf{v} \cdot \nabla)\mathbf{v} + \mathbf{f}(x,t), \\ \mathrm{div}\ \mathbf{v} = 0, \end{cases} \tag{1.5.4}$$

where $\mathbf{v}(x,t)$, $p(x,t)$ are unknown functions, i.e. for a constant μ,

$$\begin{cases} \rho\frac{\partial \mathbf{v}}{\partial t} = \mu\Delta\mathbf{v} - \rho(\mathbf{v} \cdot \nabla)\mathbf{v} - \nabla p + \mathbf{f}(x,t), \\ \mathrm{div}\ \mathbf{v} = 0. \end{cases} \tag{1.5.5}$$

This equation is called the *Navier–Stokes equation*. If this equation is set in a bounded domain, then it needs an initial condition,

$$\mathbf{v}\ |_{t=0} = \boldsymbol{\varphi}(x), \tag{1.5.6}$$

and a boundary condition, normally,

$$\mathbf{v}\ |_{\partial G} = \boldsymbol{\psi}(x,t), \tag{1.5.7}$$

where $\boldsymbol{\varphi}$ and $\boldsymbol{\psi}$ are given functions such that

$$\int_{\partial G} \boldsymbol{\psi} \cdot \mathbf{n}ds = 0.$$

If we neglect the term $\rho(\mathbf{v} \cdot \nabla)\mathbf{v}$, we obtain a linearized version called *Stokes equation*:

$$\begin{cases} \rho\frac{\partial \mathbf{v}}{\partial t} = \mu\Delta\mathbf{v} - \nabla p + \mathbf{f}(x,t), \\ \mathrm{div}\ \mathbf{v} = 0. \end{cases} \tag{1.5.8}$$

We can consider the 2-D version where $x = (x_1, x_2)$. As an eventual generalization, we can consider non-Newtonian fluids where μ can depend on $D(\mathbf{v})$.

6.   Coupling of Different Models

Models considered in the previous sections may be coupled; for instance, the heat equation and the elasticity equation form a model for thermoelasticity. The coupling may be provided in different ways. One of the possibilities is the influence of temperature on thermal dilation of a material, where the force $\mathbf{f}$ on the right-hand side of (1.4.5) is related to the gradient of the temperature:

$$\mathbf{f} = \nabla(-\beta T), \qquad (1.6.1)$$

where T is the temperature and $\beta = (3\lambda + 2\mu)\alpha$, α is the thermal dilation coefficient. The modules a_{ij}^{kl} can also depend on the temperature.

Another example is the coupling of the diffusion equation or the heat equation with the Navier–Stokes equation, where the convective term $(\mathbf{v} \cdot \nabla)u$ contains the velocity $\mathbf{v}$ defined by the Navier–Stokes equation. The viscosity μ can also depend on the concentration of one or several substances diluted in the fluid.

The reaction–diffusion equation couples the diffusion equation with kinetics of a chemical reaction:

$$\frac{\partial c}{\partial t} = \operatorname{div}(\mathcal{D}\nabla c) + R(c) + f(x,t), \qquad (1.6.2)$$

where R is a known function.

This equation is called Kolmogorov–Petrovsky–Piskunov (KPP) equation. Let us present several forms of function R:

$$R(c) = c(1 - c), \quad R(c) = c(1 - c^2),$$
$$R(c) = c(1 - c)(c - \alpha), \quad \alpha \in (0, 1),$$
$$R(c) = c^2 - c^3.$$

In the case of multiple substances, we get an analogous system of equations:

$$\begin{cases} \frac{\partial c_i}{\partial t} = \operatorname{div}(\mathcal{D}_i \nabla c_i) + R_i(c_1, \ldots, c_N) + f_i, \\ i = 1, \ldots, N. \end{cases} \qquad (1.6.3)$$

Chapter 2

Analysis of the Main Equations of Mathematical Physics

1. Some Elements of Functional Analysis

1.1. *Vector spaces*

Let us recall several notions of analysis. A set E is called a *vector space* ($\mathbb{R}$-vector space or $\mathbb{C}$-vector space) if E is supplied with two operations. The first operation, called addition $+: E \times E \to E$, takes any two vectors (elements of the set E) $\mathbf{v}$ and $\mathbf{w}$ and assigns to them the third vector $\mathbf{v} + \mathbf{w}$, called the sum of these two vectors.

The second operation called scalar multiplication, $\cdot: \mathbb{R} \times E \to E$ or $\mathbb{C} \times E \to E$, takes any real number or complex number α and any vector $\mathbf{v}$ of E and assigns to them vector $\alpha \mathbf{v}$. These two operations adhere to eight axioms:

A1. Associativity of addition:
$$\forall \mathbf{u}, \mathbf{v}, \mathbf{w} \in E, \quad \mathbf{u} + (\mathbf{v} + \mathbf{w}) = (\mathbf{u} + \mathbf{v}) + \mathbf{w}.$$

A2. Commutativity of addition:
$$\forall \mathbf{u}, \mathbf{v} \in E, \quad \mathbf{u} + \mathbf{v} = \mathbf{v} + \mathbf{u}.$$

A3. Identity element of addition (zero vector):
$$\exists \mathbf{0}_E \in E : \forall v \in E, \quad \mathbf{v} + \mathbf{0}_E = \mathbf{0}_E + \mathbf{v} = \mathbf{v}.$$

A4. Inverse elements of addition (additive inverse):

$$\forall \mathbf{v} \in E, \exists - \mathbf{v} \in E : \mathbf{v} + (-\mathbf{v}) = \mathbf{0}_E.$$

A5. Compatibility of scalar multiplication with the field multiplication in $\mathbb{R}$ or $\mathbb{C}$:

$$\forall \alpha, \beta \in \mathbb{K}, \forall \mathbf{v} \in E, \alpha(\beta \mathbf{v}) = (\alpha \beta)\mathbf{v},$$

here, and in the following, $\mathbb{K}$ stands for $\mathbb{R}$ or $\mathbb{C}$.

A6. Identity element of scalar multiplication:

$$\forall \mathbf{v} \in E, \quad 1 \cdot \mathbf{v} = \mathbf{v}$$

(here, 1 stands for the number one).

A7. Distributivity of scalar multiplication with respect to vector addition:

$$\forall \alpha \in \mathbb{K}, \forall \mathbf{u}, \mathbf{v} \in E, \quad \alpha(\mathbf{u} + \mathbf{v}) = \alpha \mathbf{u} + \alpha \mathbf{v}.$$

A8. Distributivity of scalar multiplication with respect to field addition in $\mathbb{R}$ or $\mathbb{C}$:

$$\forall \alpha, \beta \in \mathbb{K}, \forall \mathbf{v} \in E, \quad (\alpha + \beta)\mathbf{v} = \alpha \mathbf{v} + \beta \mathbf{v}.$$

For instance, $\mathbb{R}^d$ is an $\mathbb{R}$-vector space and $\mathbb{C}^d$ is a $\mathbb{C}$-vector space.

1.2. *Normed spaces*

A vector space E is called a *normed* vector space if it is supplied with a *norm*, which is a real-valued function defined on the vector space E and has the following properties:

N1. $\forall x \in E, \| x \| \geq 0; \| x \| = 0 \Leftrightarrow x = 0_E$;

N2. $\forall \alpha \in \mathbb{K}, \forall x \in E, \| \alpha x \| = | \alpha | \| x \|$;

N3. $\forall x, y \in E, \| x + y \| \leq \| x \| + \| y \|$ (triangle inequality).

The last property yields $\left| \|x\| - \|y\| \right| \leq \|x - y\|$. As an example, one can consider the so-called Euclidean norm in $\mathbb{R}^d$:

$$\|(y_1, \ldots, y_d)\|_2 = \sqrt{|y_1|^2 + \cdots + |y_d|^2}.$$

Let $(x_n)_{n \in \mathbb{N}}$ be a sequence such that $x_n \in E$, where E is a normed space.

Definition 1.1. $(x_n)_{n \in \mathbb{N}}$ **converges** to an element $\bar{x} \in E$ (denoted by $x_n \to \bar{x}$) if $\|x_n - \bar{x}\| \to 0$.

Let us formulate some direct consequences of the corresponding assertions for the *convergence* of numerical sequences in $\mathbb{R}$.

Theorem 1.1. *If* $(x_n)_{n \in \mathbb{N}}$ *converges to* $\bar{x}$, *then all subsequences* $(x_{k_n})_{n \in \mathbb{N}}$ *converge to the same limit* $\bar{x}$.

Theorem 1.2. *If* $(x_n)_{n \in \mathbb{N}}$ *converges to some limit, then its limit is unique.*

Theorem 1.3. *If* $(x_n)_{n \in \mathbb{N}}$ *converges, then it is bounded, i.e. the real-valued sequence* $(||x_n||)_{n \in \mathbb{N}}$ *is bounded.*

Theorem 1.4. *If* $x_n \to \bar{x}$, *then* $||x_n|| \to ||\bar{x}||$.

Proof. $\big|||x_n|| - ||\bar{x}||\big| \le ||x_n - \bar{x}|| \to 0.$

Let E be a normed space and M, N be two sets in E. M is called dense in N iff $\bar{M} \supset N$. Recall that $\bar{M}$ is the closure of M defined as in Section 2 of Chapter 1 replacing $\mathbb{R}^d$ by E. M is said to be nowhere dense in E iff it is not dense in any ball. $\qquad\qquad\square$

Definition 1.2. A sequence $(x_n)_{n \in \mathbb{N}}$ is called a *Cauchy sequence* if

$$\forall \varepsilon > 0, \ \ \exists n_0 \in \mathbb{N} \text{ such that } \forall n, m \ge n_0, ||x_n - x_m|| < \varepsilon.$$

Clearly, any convergent sequence is a Cauchy sequence because if $x_n \to \bar{x}$, then

$$\forall \varepsilon > 0, \ \ \exists n_0 \in \mathbb{N} \text{ such that } \forall n \ge n_0, ||x_n - \bar{x}|| < \varepsilon/2,$$

and so $\forall n, m \ge n_0$,

$$||x_n - x_m|| \le ||x_n - \bar{x}|| + ||x_m - \bar{x}|| < \varepsilon/2 + \varepsilon/2 = \varepsilon.$$

Definition 1.3. A normed space E is called a *complete space* if every Cauchy sequence converges to some limit belonging to E.

Examples: $\mathbb{R}$ is complete. $\mathbb{Q}$ is not complete ($\mathbb{Q}$ is the field of rational numbers, $\{\frac{m}{l} | m, l \in \mathbb{Z}, l \ne 0\}$) because a sequence $(1 + \frac{1}{n})^n \to e$, where the terms $(1 + \frac{1}{n})^n$ are rational numbers while e is irrational. So, this sequence is a Cauchy sequence, but its limit e is outside of the space $\mathbb{Q}$, and so it is not a convergent sequence in $\mathbb{Q}$. This example shows that if certain Cauchy sequences have no limit, we can "add" these missing limits and obtain a complete space: adding all irrational limits to $\mathbb{Q}$, we pass to $\mathbb{R}$, which is complete. The only problem is: how could we define these missing limits "living in $\mathbb{Q}$" so that we *a priori* don't know what is an irrational number.

In order to define these missing limits staying inside an incomplete normed space E, we will use a special algorithm called *completion* of E. Namely, we say that two Cauchy sequences $(x_n)_{n\in\mathbb{N}}$ and $(y_n)_{n\in\mathbb{N}}$ belong to the same equivalence class $\tilde{x}$ if $\lim\limits_{n\to\infty} ||x_n - y_n|| = 0$. Then, the set of all Cauchy sequences is presented as a partition of the equivalence classes: every Cauchy sequence belongs to some equivalence class, and it cannot belong to two different classes. Indeed, if $\tilde{x}$, $\tilde{y}$ are some equivalence classes and if the Cauchy sequence $(x_n)_{n\in\mathbb{N}} \in \tilde{x}$ and $(x_n)_{n\in\mathbb{N}} \in \tilde{y}$, then for **all** sequences $(x'_n)_{n\in\mathbb{N}} \in \tilde{x}$ and **all** sequences $(y'_n) \in \tilde{y}$, we have

$$||x'_n - y'_n|| \leq ||x'_n - x_n|| + ||x_n - y'_n|| \to 0.$$

So, $\tilde{x} = \tilde{y}$.

Every equivalence class can be associated to some sequence of this class, called a representant. In particular, we consider the equivalence classes $\tilde{c}$ containing constant sequences $(c)_{n\in\mathbb{N}} = (c, c, c, \ldots)$. These constant equivalence classes are in one-to-one correspondence with the elements of the normed space E.

Denote $\tilde{E}$ as the set of equivalence classes and $\tilde{E}_c$ the set of constant equivalence classes. We see that $\tilde{E}_c \sim E$ (equivalent to E).

Let us define now the operations of addition and scalar multiplication in $\tilde{E}$:

- $\tilde{x} + \tilde{y}$ is the equivalence class containing $(x_n + y_n)_{n\in\mathbb{N}}$, where $(x_n)_{n\in\mathbb{N}} \in \tilde{x}$ and $(y_n)_{n\in\mathbb{N}} \in \tilde{y}$.
- $\alpha\tilde{x}$ is the class containing $(\alpha x_n)_{n\in\mathbb{N}}$, where $\alpha \in \mathbb{R}$ or $\mathbb{C}$ and $(x_n)_{n\in\mathbb{N}} \in \tilde{x}$.

One can check that these operations satisfy axioms A1–A8 with $\tilde{0}$ containing the zero sequence.

Then, we introduce a norm in $\tilde{E}$: if $(x_n)_{n\in\mathbb{N}} \in \tilde{x}$, then $||\tilde{x}||_{\tilde{E}} = \lim\limits_{n\to\infty} ||x_n||$.[a] One can easily check that it satisfies the axioms N1–N3 and is stable with respect to the choice of the representant $(x_n)_{n\in\mathbb{N}}$.

---

[a]Clearly, if $(x_n)_{n\in\mathbb{N}}$ is a Cauchy sequence, then $(||x_n||)_{n\in\mathbb{N}}$ is also a Cauchy sequence in $\mathbb{R}$, and so it converges.

Indeed, if $(x_n)_{n \in \mathbb{N}}$ and $(x'_n)_{n \in \mathbb{N}}$ are two representants, then

$$\lim_{n \to +\infty} ||x_n|| = \lim_{n \to +\infty} ||x'_n + (x_n - x'_n)|| = \lim_{n \to +\infty} ||x'_n||.$$

It can be proved that $\tilde{E}$ is complete, that $\tilde{E}_c$ (a "copy" of E) is dense in $\tilde{E}$ and is isometric to E (so that $||\tilde{c}||_{\tilde{E}} = ||c||_E$). So, the elements of $\tilde{E}$ are considered as the "missing" limits.

Definition 1.4. A normed space E is called a *separable space* if it contains a countable dense subset.

Recall that a countable set is in one-to-one correspondence to the set $\mathbb{N}$. An example of a countable set is $\mathbb{Q}$. An example of a separable space is $\mathbb{R}$.

Note that a complete normed space is called a *Banach space*.

1.3. *Inner product spaces*

Definition 1.5. A vector space E is called an *inner product space*, or a **pre-Hilbert space**, or a Hausdorff pre-Hilbert space, if it is supplied with an *inner product* $(\cdot, \cdot)$ which is a map $E \times E \to \mathbb{K}$ ($\mathbb{K}$ is $\mathbb{R}$ or $\mathbb{C}$) satisfying the following three axioms:

I1. Linearity in the first argument:

$$\forall \alpha, \beta \in \mathbb{K}, \forall \mathbf{u}, \mathbf{v}, \mathbf{w} \in E,$$

$$(\alpha \mathbf{u} + \beta \mathbf{v}, \mathbf{w}) = \alpha(\mathbf{u}, \mathbf{w}) + \beta(\mathbf{v}, \mathbf{w}).$$

I2. Symmetry or conjugate symmetry:

$$\forall \mathbf{x}, \mathbf{y} \in E, \quad (\mathbf{x}, \mathbf{y}) = (\mathbf{y}, \mathbf{x}) \quad (\text{case } \mathbb{K} = \mathbb{R})$$

or

$$\forall \mathbf{x}, \mathbf{y} \in E, \quad (\mathbf{x}, \mathbf{y}) = (\overline{\mathbf{y}, \mathbf{x}}) \quad (\text{case } \mathbb{K} = \mathbb{C}).$$

I3. $\forall \mathbf{x} \in E$, $(\mathbf{x}, \mathbf{x})$ is real and $(\mathbf{x}, \mathbf{x}) \geq \mathbf{0}$.
$(\mathbf{x}, \mathbf{x}) = \mathbf{0}$ iff $\mathbf{x} = \mathbf{0}_E$.
An inner product space is supplied with the norm

$$||\mathbf{x}|| = \sqrt{(\mathbf{x}, \mathbf{x})}.$$

For example, $\mathbb{R}^d$ is an inner product space with $(\mathbf{x}, \mathbf{y}) = \sum_{i=1}^{d} x_i y_i$. For $\mathbb{C}^d$, the inner product is defined by the formula

$$(\mathbf{x}, \mathbf{y}) = \sum_{i=1}^{d} x_i \bar{y}_i.$$

Let us recall the main **properties** of an inner product.

P1. *Cauchy–Bunyakovsky–Schwarz (CBS) inequality:*

$$\forall \mathbf{x}, \mathbf{y} \in E, \quad |(\mathbf{x}, \mathbf{y})| \leq ||\mathbf{x}|| \, ||\mathbf{y}||.$$

Proof. For any $\lambda \in \mathbb{K}$, $\mathbf{x}, \mathbf{y} \in E$, $(\mathbf{x} - \lambda \mathbf{y}, \mathbf{x} - \lambda \mathbf{y}) \geq 0$, i.e.

$$(\mathbf{x}, \mathbf{x}) - \lambda(\mathbf{y}, \mathbf{x}) - \bar{\lambda}(\mathbf{x}, \mathbf{y}) + (\lambda \mathbf{y}, \lambda \mathbf{y}) \geq 0.$$

Setting $\lambda = \frac{(\mathbf{x}, \mathbf{y})}{(\mathbf{y}, \mathbf{y})}$ (for $\mathbf{y} \neq \mathbf{0}_E$), we get

$$(\mathbf{x}, \mathbf{x}) - \frac{(\mathbf{x}, \mathbf{y})(\overline{\mathbf{x}, \mathbf{y}})}{(\mathbf{y}, \mathbf{y})} - \frac{(\overline{\mathbf{x}, \mathbf{y}})(\mathbf{x}, \mathbf{y})}{(\mathbf{y}, \mathbf{y})} + \frac{|(\mathbf{x}, \mathbf{y})|^2}{(\mathbf{y}, \mathbf{y})} \geq 0, \text{ i.e.}$$

$$(\mathbf{x}, \mathbf{x}) - \frac{|(\mathbf{x}, \mathbf{y})|^2}{(\mathbf{y}, \mathbf{y})} \geq 0, \text{ and so}$$

$$|(\mathbf{x}, \mathbf{y})|^2 \leq ||\mathbf{x}||^2 ||\mathbf{y}||^2.$$

The case $\mathbf{y} = \mathbf{0}_E$ is evident.
The assertion is proved. $\qquad\qquad\qquad\qquad\qquad\qquad\qquad\qquad\square$

P2. *Continuity of the inner product:*
If $x_n \to \bar{x}$ and $y_n \to \bar{y}$, then

$$(x_n, y_n) \to (\bar{x}, \bar{y}).$$

Proof.

$$\begin{aligned}
\left|(x_n, y_n) - (\bar{x}, \bar{y})\right| &= \left|(x_n - \bar{x}, y_n) + (\bar{x}, y_n - \bar{y})\right| \\
&\leq \left|(x_n - \bar{x}, y_n)\right| + \left|(\bar{x}, y_n - \bar{y})\right| \\
&\leq ||x_n - \bar{x}|| \, ||y_n|| + ||\bar{x}|| \, ||y_n - \bar{y}||.
\end{aligned}$$

The right-hand side tends to zero because

$$||x_n - \bar{x}|| \to 0, \quad ||y_n|| \to ||\bar{y}||, \quad ||y_n - \bar{y}|| \to 0.$$

$\qquad\qquad\qquad\qquad\qquad\qquad\qquad\qquad\qquad\qquad\qquad\qquad\qquad\square$

P3. *Parallelogram identity:*

$$\forall \mathbf{x}, \mathbf{y} \in E, \ ||\mathbf{x} + \mathbf{y}||^2 + ||\mathbf{x} - \mathbf{y}||^2 = 2 \left(||\mathbf{x}||^2 + ||\mathbf{y}||^2 \right).$$

Proof.

$$
\begin{aligned}
||\mathbf{x} + \mathbf{y}||^2 + ||\mathbf{x} - \mathbf{y}||^2 &= (\mathbf{x} + \mathbf{y}, \mathbf{x} + \mathbf{y}) + (\mathbf{x} - \mathbf{y}, \mathbf{x} - \mathbf{y}) \\
&= (\mathbf{x}, \mathbf{x}) + (\mathbf{y}, \mathbf{x}) + (\mathbf{x}, \mathbf{y}) + (\mathbf{y}, \mathbf{y}) \\
&\quad + (\mathbf{x}, \mathbf{x}) - (\mathbf{y}, \mathbf{x}) - (\mathbf{x}, \mathbf{y}) + (\mathbf{y}, \mathbf{y}) \\
&= 2||\mathbf{x}||^2 + 2||\mathbf{y}||^2.
\end{aligned}
$$
$\square$

The completion of an inner product space is compatible with the inner product

$$(\tilde{x}, \tilde{y})_{\tilde{E}} = \lim_{n \to +\infty} (x_n, y_n),$$

where $(x_n)_{n \in \mathbb{N}}$, $(y_n)_{n \in \mathbb{N}}$ are representants of $\tilde{x}$ and $\tilde{y}$, respectively.

Definition 1.6. A complete inner product space E is called a *Hilbert space*. We will consider separable Hilbert spaces.

1.4. Linear operators

Let us introduce linear operators. Let E, F be normed spaces. A mapping $A : E \to F$ is called a *linear operator* if $\forall \alpha, \beta \in \mathbb{K}, \mathbf{u}, \mathbf{v} \in E$,

$$A(\alpha \mathbf{u} + \beta \mathbf{v}) = \alpha A \mathbf{u} + \beta A \mathbf{v}.$$

The linear operator A is called a *bounded operator* if

$$\exists M \geq 0, \ \forall \mathbf{u} \in E, \ ||A\mathbf{u}||_F \leq M||\mathbf{u}||_E.$$

The linear operator A is called a *continuous operator* if for any convergent sequence $(\mathbf{x}_n)_{n \in \mathbb{N}}$ of the space E, the sequence of the images $(A\mathbf{x}_n)_{n \in \mathbb{N}}$ converges to $A\mathbf{x}_0$, where $\mathbf{x}_0$ is the limit $(\mathbf{x}_n)_{n \in \mathbb{N}}$.

Theorem 1.5. *A is continuous iff it is bounded.*

Proof.

1. Assume that A is **bounded**. Let $(\mathbf{x}_n)_{n \in \mathbb{N}}$ be a sequence converging to $\mathbf{x}_0$. So, $\exists M \geq 0$ such that $||A(\mathbf{x}_n - \mathbf{x}_0)||_F \leq M||\mathbf{x}_n - \mathbf{x}_0||_E$. Taking into account that $||\mathbf{x}_n - \mathbf{x}_0||_E \to 0$, we derive that $A\mathbf{x}_n \to A\mathbf{x}_0$. So, A is **continuous**.

2. Assume now that A is continuous. If A is not bounded, then there exists a sequence $(\mathbf{x}_n)_{n\in\mathbb{N}}$ in E such that $||\mathbf{x}_n||_E = 1$ and $||A\mathbf{x}_n|| > n$.

Consider the sequence $(\frac{1}{n}\mathbf{x}_n)_{n\in\mathbb{N}} \to \mathbf{0}_E$. However, $||A\left(\frac{1}{n}\mathbf{x}_n\right)||_F > \frac{n}{n} = 1$. So, $A(\frac{1}{n}\mathbf{x}) \not\to A\mathbf{0}_E = \mathbf{0}_E$. It is incompatible with the continuity of A. So, A cannot be unbounded. The theorem is proved. $\quad\square$

If A is a bounded operator, we can define its norm as

$$||A||_{\mathcal{L}(E,F)} = \sup_{\mathbf{x}\neq\mathbf{0}_E}\frac{||A\mathbf{x}||_F}{||\mathbf{x}||_E} = \sup_{\mathbf{x}:||\mathbf{x}||_E=1}||A\mathbf{x}||_F.$$

One can easily check that it satisfies axioms N1–N3.

If $F = \mathbb{R}$ or $\mathbb{C}$, then A is called a *linear form* or *linear functional*.

Theorem 1.6. (*Riesz–Fréchet representation theorem*) *Let H be a Hilbert space. For every continuous linear functional $\varphi : H \to \mathbb{K}$, there exists a unique element $\mathbf{y} \in H$ such that for all $\mathbf{x} \in H$,*

$$\varphi(\mathbf{x}) = (\mathbf{x}, \mathbf{y}),$$

and moreover

$$||\mathbf{y}||_H = ||\varphi||_{\mathcal{L}(H,\mathbb{K})}.$$

Proof. See Appendix B and Ref. [1]. $\quad\square$

The space of continuous linear functionals is called **dual to H** and is denoted by H^*.

Definition 1.7. A sequence $(\mathbf{x}_n) \in H$ is called *weakly convergent* in H to the element $\mathbf{x}_0$ if for all continuous linear functionals $\varphi \in H^*$,

$$\varphi(\mathbf{x}_n) \to \varphi(\mathbf{x}_0).$$

This means that for all elements $\mathbf{y} \in H$,

$$(\mathbf{x}_n, \mathbf{y}) \to (\mathbf{x}_0, \mathbf{y}).$$

Clearly, every convergent sequence weakly converges to its limit. The inverse assertion is not true.

Theorem 1.7. (*Banach–Alaoglu theorem* **for a separable Hilbert space**) *The closed unit ball $\overline{B(\mathbf{0}_H, 1)}$ in a separable Hilbert space H is relatively weakly compact. This means that every sequence in $\overline{B(\mathbf{0}_H, 1)}$ has a weakly convergent subsequence.*

Proof.   See Ref. [1].   □

2.   Sobolev Spaces

2.1.   *Auxiliary spaces*

Let G be a bounded domain in $\mathbb{R}^d$. Let us introduce some vector spaces:

(1) $C(\bar{G})$ is the set of continuous functions $\bar{G} \to \mathbb{R}$ (extendable to $C(\bar{B})$, where B is a ball in $\mathbb{R}^d$ containing $\bar{G}$).
(2) $C^{(k)}(\bar{G})$ is the set of k times continuously differentiable functions $\bar{G} \to \mathbb{R}$, also extendable to $C^{(k)}(\bar{B})$.

Remark 2.1. For $G \subset B$, $\tilde{u}: B \to \mathbb{R}$ is an *extension* of a function u: $\bar{G} \to \mathbb{R}$ if $\forall x \in \bar{G}$, $\tilde{u}(x) = u(x)$. In this case, u is called a *restriction* of $\tilde{u}$.

(3) $C_0(G)$, $C_0^{(k)}(G)$ are the subspaces of $C(\bar{G})$ and $C^{(k)}(\bar{G})$ such that their functions vanish in some neighborhood G_ε of the boundary ∂G; more exactly, f belongs to these subspaces if there exists a positive ε such that f vanishes in

$$G_\varepsilon = \{x \in G | \mathrm{dist}(x, \partial G) < \varepsilon\}.$$

Such a function is called a *compact support function*, i.e. $\overline{\mathrm{supp}\, f} \subset G$, where $\mathrm{supp}\, f = \{x \in G | f(x) \neq 0\}$. Here, dist (x, A) is a distance between the point x and the set A. It is defined as follows:

$$\mathrm{dist}\,(x, A) = \inf_{y \in A} \|x - y\|_2.$$

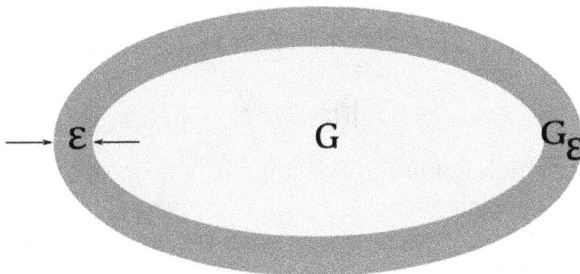

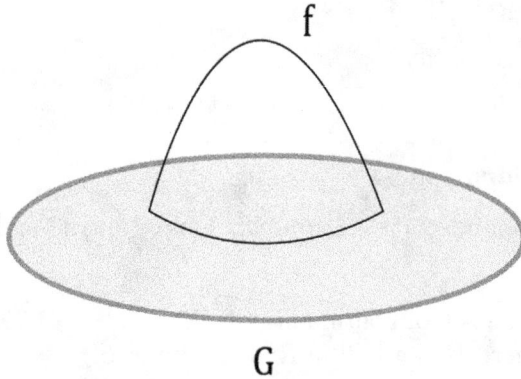

Let us recall the definition of L^p space, $1 \le p < +\infty$ (*Lebesgue spaces*). For a bounded domain G, $L^p(G)$ is the vector space of functions $f : G \to \mathbb{R}$ having finite Lebesgue's integral

$$\int_G |f(x)|^p dx.$$

$L^p(G)$ is supplied with the norm

$$\|f\|_{L^p(G)} = \left(\int_G |f(x)|^p dx \right)^{1/p}. \tag{2.2.1}$$

$L^p(G)$ can be equivalently defined as a completion of $C(\bar{G})$ or $C_0^{(\infty)}(G)$ with respect to the norm (2.2.1).

So, $L^p(G)$ is a Banach space such that every function $f \in L^p(G)$ is a limit of a sequence $(f_n)_{n\in\mathbb{N}}$, where $f_n \in C(\bar{G})$ or $C_0^{(\infty)}(G)$.

According to the Weierstrass theorem, every continuous function is the limit of a sequence of polynomials in the sense of $\|\cdot\|_{C(\bar{G})}$ norm, and so also in the sense of L^p norm, because

$$\|f\|_{L^p(G)} = \left(\int_G |f(x)|^p dx \right)^{1/p} \le \left(\int_G \max_{x\in\bar{G}} |f(x)|^p dy \right)^{1/p}$$

$$= \max_{x\in\bar{G}} |f(x)| (\mathrm{mes}G)^{1/p} = \|f\|_{C(\bar{G})} (\mathrm{mes}G)^{1/p}.$$

Also, every polynomial is the limit of a sequence of polynomials with rational coefficients.

So, the set $\mathbb{Q}[X]$ of polynomials with rational coefficients is dense in L^p. Thus, $L^p(G)$ is a separable space.

If $p = 2$, we can introduce an inner product in $L^2(G)$:

$$(f, g) = \int_G f(x)g(x)dx, \qquad (2.2.2)$$

so $L^2(G)$ is a separable Hilbert space.

Finally, let us define $L^\infty(G)$ as the space of functions $f : G \to \mathbb{R}$ such that $\exists M \geq 0$ such that $|f(x)| \leq M$ almost everywhere, i.e. for all $x \in G$ except for a subset E having measure zero. This space is supplied with the L^∞ norm:

$$\|f\|_{L^\infty(G)} = \text{vrai} \max_{x \in G} |f(x)|,$$

also called $\operatorname{ess\,sup}_{x \in G} |f(x)|$ which is

$$\inf_{E \subset G, \text{mes } E = 0} \left\{ \sup_{x \in G \backslash E} |f(x)| \right\}.$$

$L^\infty(G)$ is a non-separable Banach space.

2.2.   *Sobolev space H^1*

Consider a bounded domain G with *Lipschitz boundary* ∂G. This means that ∂G can be covered by a finite set of open balls $O_1, \ldots, O_N$ such that in each O_i in some local coordinate system obtained from the original one by rotations, $\partial G \cap O_i$ is a graph of a Lipschitz function. (f is called *L-Lipschitz* on G if $\forall x, y \in G$, $|f(x) - f(y)| \leq L\|x - y\|_2$.)

Similarly, we define the $C^{(k)}$-smooth boundary.

Definition 2.1. The *Sobolev space* $H^1(G)$ is a completion of $C^{(\infty)}(\bar{G})$ with respect to the inner product

$$(f, g) = \int_G (f(x)g(x) + \nabla f(x) \cdot \nabla g(x)) \, dx. \qquad (2.2.3)$$

Arguing as for L^p, we can prove that $H^1(G)$ is a separable Hilbert space. Note that in the definition, $C^{(\infty)}(\bar{G})$ can be replaced by

$C^{(1)}(\bar{G})$. The norm in $H^1(G)$ is defined as

$$\|f\|_{H^1(G)} = \sqrt{(f,f)} = \sqrt{\int_G |f(x)|^2 + \|\nabla f\|_2^2 dx}$$

$$= \sqrt{\|f\|_{L^2(G)}^2 + \|\nabla f\|_{L^2(G)}^2}.$$

So, $\|f\|_{H^1(G)} \geq \|f\|_{L^2(G)}$.

Definition 2.2. $H_0^1(G)$ is a completion of $C_0^{(\infty)}(G)$ with respect to the inner product (2.2.3).

So, $H_0^1(G)$ is a subspace of $H^1(G)$ "vanishing at the boundary". Note that Definition 2.2 is still valid for any bounded domain G.

Let f be a function of $H^1(G)$. Then, there exists a sequence $f_n \in C^{(\infty)}(\bar{G})$ convergent to f (because $C^{(\infty)}(\bar{G})$ is dense in $H^1(G)$). This sequence is a Cauchy sequence in $H^1(G)$, and so $\forall \varepsilon > 0 \exists n_0 \in \mathbb{N}$ such that $\forall n, m \geq n_0$,

$$\sqrt{\|f_n - f_m\|_{L^2(G)}^2 + \sum_{i=1}^d \|\frac{\partial f_n}{\partial x_i} - \frac{\partial f_m}{\partial x_i}\|_{L^2(G)}^2} < \varepsilon.$$

So, $(f_n)_{n \in \mathbb{N}}$ is a Cauchy sequence in $L^2(G)$ and $(\frac{\partial f_n}{\partial x_i})_{n \in \mathbb{N}}$ for all $i = 1, \ldots, d$ are Cauchy sequences in $L^2(G)$ as well . As $L^2(G)$ is complete, these sequences have limits: $f_n \to f$, $\frac{\partial f_n}{\partial x_i} \to w_i$, $f, w_i \in L^2(G)$. The functions w_i are called *weak partial derivatives* of f, denoted by $\frac{\partial f}{\partial x_i}$.

Theorem 2.1. $\forall f \in H^1(G)$, $v \in C_0^{(\infty)}(G)$, $\forall i = 1, \ldots, d$,

$$\int_G f(x) \frac{\partial v}{\partial x_i}(x) dx = -\int_G \frac{\partial f}{\partial x_i}(x) v(x) dx. \tag{2.2.4}$$

Proof. Let $(f_n)_{n \in \mathbb{N}}$ be a sequence convergent to f in $H^1(G)$. For every n, integrating by parts and taking into consideration that v vanishes in the neighborhood of ∂G, we get

$$\int_G f_n(x) \frac{\partial v}{\partial x_i}(x) dx = -\int_G \frac{\partial f_n}{\partial x_i}(x) v(x) dx.$$

Passing to the limit in this equality, we get (2.2.4). $\qquad\square$

Remark 2.2. Let v be a function of $H_0^1(G)$. Then, considering a sequence $(v_n)_{n \in \mathbb{N}}$ in $C_0^{(\infty)}(G)$ convergent to v and passing to the limit, we prove that (2.2.4) is still valid for $v \in H_0^1(G)$.

Theorem 2.2. *If* $f \in H^1(G)$ *and* $w_i \in L^2(G)$ *such that* $\forall v \in H_0^1(G)$,

$$\int_G f(x) \frac{\partial v}{\partial x_i} dx = - \int_G w_i(x) v(x) dx.$$

Then, $w_i = \frac{\partial f}{\partial x_i}$ *(so,* $\frac{\partial f}{\partial x_i}$ *is uniquely defined).*

Proof. Theorem 2.1 yields

$$\int_G f(x) \frac{\partial v}{\partial x_i}(x) dx = - \int_G \frac{\partial f}{\partial x_i} v(x) dx.$$

So,

$$\int_G \left(w_i - \frac{\partial f}{\partial x_i} \right) v(x) dx = 0$$

for all $v \in H_0^1(G)$. $H_0^1(G)$ is dense in $L^2(G)$ because $C_0^{(\infty)}(G)$ is dense in $L^2(G)$. So, $w_i - \frac{\partial f}{\partial x_i}$ is orthogonal to all functions of $L^2(G)$. So, $w_i - \frac{\partial f}{\partial x_i} = 0$. $\qquad \square$

Comment. Theorems 2.1 and 2.2 show that the definition of the weak partial derivatives $\frac{\partial f}{\partial x_i}$ is equivalent to the following one: $w_i \in L^2(G)$ is a weak partial derivative $\frac{\partial f}{\partial x_i}$ of the function $u \in L^2(G)$ if and only if $\forall v \in C_0^{(\infty)}(G)$,

$$\int_G f(x) \frac{\partial v}{\partial x_i}(x) dx = - \int_G w_i v(x) dx.$$

This definition allows us to introduce the Sobolev space $H^1(G)$ as the space of functions of $L^2(G)$ such that for all $i = 1, \ldots, d$, they have weak partial derivatives $\partial/\partial x_i$ belonging to $L^2(G)$.

For the functions of $H^1(G)$ with Lipschitz ∂G, one can introduce the notion of the *trace of a function* on ∂G. Let us show it in the case of the G-parallelepiped

$$\Pi = \{ x = (x_1, \ldots, x_d) | 0 < x_i < l_i, \ i = 1, \ldots, d \}.$$

Denote $x' = (x_2, \ldots, x_d)$,

$$\Gamma = \{x_1 = 0, \ 0 < x_i < l_i, \ i = 2, \ldots, d\}$$

as a part of ∂G. Let u be a function of $H^1(\Pi)$, and define the trace of u on Γ.

Lemma 2.1. *Let* $u \in C^{(\infty)}(\bar{\Pi})$*, then* $\forall \delta > 0$*,* $\delta < l_1$*,*

$$\int_\Gamma u^2(0, x') dx' \leq \frac{2}{\delta} \int_{Q_\delta(\Gamma)} u^2(x) dx + \delta \int_{Q_\delta(\Gamma)} (\partial u / \partial x_1)^2 dx,$$

$$Q_\delta(\Gamma) = \{x \in \Pi | x_1 \in (0, \delta)\}.$$

Proof.

$$\int_\Gamma u^2(0, x') dx' = \frac{1}{\delta} \int_{Q_\delta(\Gamma)} \left[-u(x) + \int_0^{x_1} \frac{\partial u(\tau, x')}{\partial \tau} d\tau \right]^2 dx$$

$$\leq \frac{2}{\delta} \int_{Q_\delta(\Gamma)} u^2(x) dx + \frac{2}{\delta} \int_\Gamma \left\{ \int_0^\delta \left(\int_0^{x_1} \frac{\partial u(\tau, x')}{\partial \tau} d\tau \right)^2 dx_1 \right\} dx'$$

$$\leq \frac{2}{\delta} \int_{Q_\delta(\Gamma)} u^2(x) dx + \frac{2}{\delta} \int_\Gamma \left\{ \int_0^\delta x_1 \int_0^\delta \left(\frac{\partial u(\tau, x')}{\partial \tau} \right)^2 d\tau dx_1 \right\} dx',$$

where we used the CBS inequality

$$\left(\int_0^{x_1} \varphi(\tau) d\tau \right)^2 = \left(\int_0^{x_1} 1 \varphi(\tau) d\tau \right)^2 \leq \int_0^{x_1} 1^2 d\tau \cdot \int_0^{x_1} (\varphi(\tau))^2 d\tau$$

$$\leq x_1 \int_0^\delta (\varphi(\tau))^2 d\tau,$$

and Young's inequality $(A + B)^2 \leq 2A^2 + 2B^2$.

So,

$$\int_\Gamma u^2(0, x') dx' \leq \frac{2}{\delta} \int_{Q_\delta(\Gamma)} u^2(x) dx + \delta \int_{Q_\delta(\Gamma)} \left(\frac{\partial u}{\partial x_1} \right)^2 dx. \qquad \square$$

This lemma shows that every Cauchy sequence $(u_n)_{n \in \mathbb{N}} \in C^{(\infty)}(\bar{\Pi})$ generates the sequence of traces $(u_n(0, x'))_{n \in \mathbb{N}}$ which is a Cauchy sequence in $L^2(\Gamma)$, and so it converges to some function of $L^2(\Gamma)$.

Now, let $(u_n)_{n \in \mathbb{N}}$ be convergent to $u \in H^1(\Pi)$, and denote $u|_\Gamma$ as a function of $L^2(\Gamma)$ which is the limit of the Cauchy sequence of traces $(u_n(0, x'))_{n \in \mathbb{N}}$. This function $u|_\Gamma$ is called the trace of the function u.

3. Poincaré's Inequalities

Let G be a bounded domain in $\mathbb{R}^d$.

3.1. *Poincaré–Friedrichs inequality*

Theorem 3.1. (*Poincaré–Friedrichs inequality*) *There exists* $C_{\mathrm{PF}} \geq 0$ *such that* $\forall u \in H_0^1(G)$.

$$\|u\|_{L^2(G)} \leq C_{\mathrm{PF}} \|\nabla u\|_{L^2(G)}, \tag{2.3.1}$$

where $\|\nabla u\|_{L^2(G)} = \sqrt{\int_G \sum_{i=1}^d (\frac{\partial u}{\partial x_i})^2 dx}$ *and*

$$C_{\mathrm{PF}} = \frac{\mathrm{diam}\ G}{\sqrt{2}}$$

(diam $G = \sup_{x,y \in G} |x - y|$). *This inequality is called Poincaré–Friedrichs (PF) inequality.*

Proof. (1) Let $\Pi \supset G$ be the parallelepiped

$$\Pi = (a_1, b_1) \times \cdots \times (a_d, b_d) \ \text{ and } \ \forall i = 1, \ldots, d, \ b_i - a_i \leq \mathrm{diam}\ G.$$

Without loss of generality, assume $a_1 = 0$.

Consider $u \in C_0^{(\infty)}(\bar{\Pi})$. Then,

$$u(x_1, x_2, \ldots, x_d) = \int_0^{x_1} \frac{\partial u}{\partial \tau}(\tau, x') d\tau, \ \ x' = (x_2, \ldots, x_n).$$

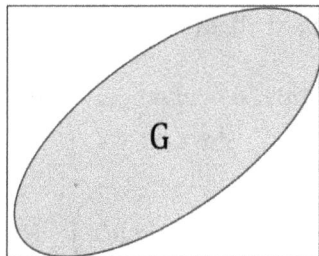

Π

So, applying the CBS inequality, we obtain

$$\|u\|_{L^2(\Pi)}^2 = \int_\Pi \left(\int_0^{x_1} \frac{\partial u}{\partial \tau}(\tau, x')d\tau \right)^2 dx$$

$$\leq \int_\Pi \left\{ \int_0^{x_1} 1^2 dx_1 \int_0^{x_1} \left(\frac{\partial u}{\partial \tau}(\tau, x') \right)^2 d\tau \right\} dx_1 dx'$$

$$\leq \int_{a_2}^{b_2} \cdots \int_{a_d}^{b_d} \int_{a_1}^{b_1} x_1 \int_{a_1}^{b_1} \left(\frac{\partial u}{\partial \tau}(\tau, x') \right)^2 d\tau dx_1 dx'$$

$$= \frac{(b_1 - a_1)^2}{2} \int_\Pi \left(\frac{\partial u}{\partial x_1} \right)^2 dx \leq C_{\mathrm{PF}}^2 \|\nabla u\|_{L^2(\Pi)}^2. \quad (2.3.2)$$

(2) Let u be a function of $C_0^{(\infty)}(G)$. Extend it by zero to $\Pi \backslash G$. The extension is $\tilde{u} \in C_0^{(\infty)}(\Pi)$ and for this, we obtain (2.3.2). On the other hand, $\|\tilde{u}\|_{L^2(\Pi)} = \|u\|_{L^2(G)}$ and $\|\nabla \tilde{u}\|_{L^2(\Pi)} = \|\nabla u\|_{L^2(G)}$. So, we get (2.3.1) for $u \in C_0^{(\infty)}(G)$.

(3) Let u be a function of $H_0^1(G)$. Let $(u_n)_{n\in\mathbb{N}}$ be a sequence of functions of $C_0^{(\infty)}(G)$ convergent to u; for all n,

$$\|u_n\|_{L^2(G)} \leq C_{\mathrm{PF}} \|\nabla u_n\|_{L^2(G)}.$$

Passing to the limit and using the continuity of a norm, we get (2.3.1).
The theorem is proved.   □

3.2.   *Poincaré's inequality in a parallelepiped*

Let G be a parallelepiped $(0, a_1) \times \cdots \times (0, a_d) \subset \mathbb{R}^d$.

Theorem 3.2. (*Poincaré's inequality*) *There exists a positive constant C_P such that $\forall u \in H^1(G)$,*

$$\|u\|_{L^2(G)}^2 \leq \frac{1}{\mathrm{mes}(G)} \left(\int_G u(x)dx \right)^2 + C_P \|\nabla u\|_{L^2(G)}^2.$$

Proof.

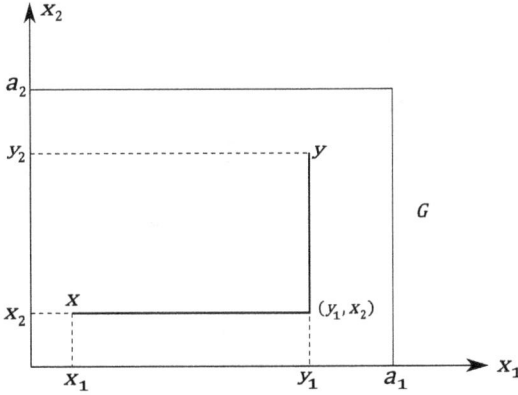

1. Consider $u \in C^{(\infty)}(\bar{G})$, $x, y \in G$. Then,

$$(u(y) - u(x))^2$$

$$= \left(\int_{x_1}^{y_1} \frac{\partial u}{\partial \theta_1}(\theta_1, x_2, \ldots, x_d) d\theta_1 \right.$$

$$\left. + \cdots + \int_{x_d}^{y_d} \frac{\partial u}{\partial \theta_d}(y_1, \ldots, y_{d-1}, \theta_d) d\theta_d \right)^2$$

$$\leq d \left\{ \left(\int_{x_1}^{y_1} \frac{\partial u}{\partial \theta_1} d\theta_1 \right)^2 + \cdots + \left(\int_{x_d}^{y_d} \frac{\partial u}{\partial \theta_d} d\theta_d \right)^2 \right\}$$

(by Cauchy–Schwarz inequality, $(a_1 + \cdots + a_d)^2 \leq d(a_1^2 + \cdots + a_d^2)$), and so it is smaller than (CSB)

$$d \left\{ a_1 \int_0^{a_1} \left(\frac{\partial u}{\partial \theta_1} \right)^2 d\theta_1 + \cdots + a_d \int_0^{a_d} \left(\frac{\partial u}{\partial \theta_d} \right)^2 d\theta_d \right\}.$$

Integrating the inequality in $x, y \in G$, we get

$$\int_G \int_G (u(y) - u(x))^2 dx dy$$

$$\leq d \left\{ a_1^2 \text{ mes } G \int_G \left(\frac{\partial u}{\partial x_1}(x) \right)^2 dx \right.$$

$$\left. + \cdots + a_d^2 \text{ mes } G \int_G \left(\frac{\partial u}{\partial x_d}(x) \right)^2 dx \right\}$$

$$\leq d \text{ mes } G \max_{1 \leq j \leq d} a_j^2 \|\nabla u\|_{L^2(G)}^2.$$

The left-hand side is

$$\int_G \int_G (u(y))^2 dx dy - 2 \int_G \int_G u(y)u(x) dx dy + \int_G \int_G (u(x))^2 dx dy$$

$$= 2 \text{ mes } G \|u\|_{L^2(G)}^2 - 2 \left(\int_G u(x) dx \right)^2.$$

Finally, we get

$$\|u\|_{L^2(G)}^2 \leq \frac{1}{\text{mes } G} \left(\int_G u(x) dx \right)^2 + \frac{d}{2} \max_{1 \leq j \leq d} a_j^2 \|\nabla u\|_{L^2(G)}^2.$$

So, $C_P = \frac{d}{2} \max_{1 \leq j \leq d} a_j^2$.

2. Consider $u \in H^1(G)$. Let $(u_n)_{n \in \mathbb{N}}$ be a sequence such that $u_n \in C^{(\infty)}(\bar{G})$ and $u_n \to u$ in H^1 norm. Then, we get

$$\|u_n\|_{L^2(G)} \to \|u\|_{L^2(G)}, \quad \|\nabla u_n\|_{L^2(G)} \to \|\nabla u\|_{L^2(G)}.$$

In the first part of the proof, we have obtained

$$\|u_n\|_{L^2(G)}^2 \leq \frac{1}{\text{mes }(G)} \left(\int_G u_n(x) dx \right)^2 + C_P \|\nabla u_n\|_{L^2(G)}^2.$$

The continuity of a norm yields

$$\|u_n - u\|_{L^2(G)}^2 + \|\nabla u_n - \nabla u\|_{L^2(G)}^2 = \|u_n - u\|_{H^1(G)}^2 \to 0,$$

and so

$$\left| \int_G u_n dx - \int_G u dx \right| = \left| \int_G (u_n - u) 1 dx \right|$$

$$\leq \sqrt{\int_G 1^2 dx} \|u_n - u\|_{L^2(G)} \to 0.$$

So, finally, $\int_G u_n dx$ also converges to $\int_G u(x) dx$.

The theorem is proved. It can be generalized for arbitrary bounded domain with Lipschitz boundary. $\qquad \square$

4. Stationary Conductivity Equation

Consider the following problem:

$$-\operatorname{div}(A(x)\nabla u) = f(x), \quad x \in G, \tag{2.4.1}$$

$$u\big|_{\partial G} = 0, \tag{2.4.2}$$

where G is a bounded domain in $\mathbb{R}^d$, f is a function of $L^2(G)$, and

$$A(x) = (A_{ij}(x))_{1 \leq i,j \leq d}$$

is $d \times d$ symmetric, positive definite matrix, satisfying:

(i) $\forall x \in G, \ \forall i,j \in \{1,\ldots,d\}, \ A_{ij}(x) = A_{ji}(x);$
(ii) $\exists \kappa > 0 : \forall x \in G, \ \forall \boldsymbol{\xi} = (\xi_1,\ldots,\xi_d),$

$$\sum_{i,j=1}^{d} A_{ij}(x)\xi_j\xi_i \geq \kappa \sum_{i=1}^{d} \xi_i^2.$$

Currently, we assume that $A_{ij} \in C^{(1)}(\bar{G})$; however, in the following, we will define a weak solution, and it will be valid for any bounded measurable function A_{ij}.

Namely, the notion of the classical solution, satisfying equation (2.4.1) in each point $x \in G$ and vanishing on ∂G, can be generalized for the case of non-smooth coefficients by means of a **weak formulation**.

The function $u \in H_0^1(G)$ is called a **weak solution** of problem (2.4.1), (2.4.2), if $\forall v \in C_0^{(\infty)}(G)$, the following integral identity holds:

$$\int_G A(x)\nabla u \cdot \nabla v \, dx = \int_G f(x)v(x)\,dx. \tag{2.4.3}$$

Evidently, if A_{ij} are smooth functions and u is a classical solution, then multiplying equation (2.4.1) by a test function $v \in C_0^{(\infty)}(G)$, we integrate it over G; integrate the left-hand side by parts and obtain (2.4.3).

Note that (2.4.1) can be rewritten as

$$-\sum_{i,j=1}^{d} \frac{\partial}{\partial x_i}\left(A_{ij}(x)\frac{\partial u}{\partial x_j}\right) = f(x), \qquad (2.4.1')$$

and (2.4.3) as

$$\int_G \sum_{i,j=1}^{d} A_{ij}(x)\frac{\partial u}{\partial x_j}\frac{\partial v}{\partial x_i}dx = \int_G f(x)v(x)dx. \qquad (2.4.3')$$

Note that due to the density of $C_0^{(\infty)}(G)$ in $H_0^1(G)$, (2.4.3′) is valid $\forall v \in H_0^1(G)$.

So, a weak solution is more general: it can be considered in the case of discontinuous coefficients.

Let us prove the following theorem.

Theorem 4.1. *For any function $f \in L^2(G)$, a weak solution u exists, is unique and satisfies the inequality*

$$\|u\|_{H^1(G)} \leq C_D\|f\|_{L^2(G)}, \qquad (2.4.4)$$

where the constant

$$C_D = \frac{\sqrt{2}}{\kappa}\max\{C_{PF}, C_{PF}^2\}. \qquad (2.4.5)$$

Proof.

1. Consider the space $H_0^1(G)$ supplied with a new inner product

$$[u, v] = \int_G \sum_{i,j=1}^{d} A_{ij}(x)\frac{\partial u}{\partial x_j}\frac{\partial v}{\partial x_i}dx.$$

It is easy to check its linearity with respect to the first argument and symmetry. Moreover, $[v, v] \geq 0$ for all $v \in H_0^1(G)$ because of the property **(ii)** of coefficients A_{ij}. If $[v, v] = 0$, then due to **(ii)** $\|\nabla v\|_{L^2(G)}^2 = 0$, and then due to the PF inequality, $\|v\|_{L^2(G)}^2 = 0$. So, $v = 0$.

The norm $[|v|] = \sqrt{[v, v]}$ is equivalent to the norm $H^1(G)$, i.e. there exist constants $C_1, C_2 > 0$ such that for all $v \in H_0^1(G)$,

$$C_1\|v\|_{H^1(G)} \leq [|v|] \leq C_2\|v\|_{H^1(G)}.$$

Denote H as the space $H_0^1(G)$ supplied with the inner product $[u, v]$.

2. Consider the linear functional $\Phi\colon H \to \mathbb{R}$ such that

$$\Phi(v) = \int_G f(x)v(x)dx.$$

Clearly, $\forall \alpha,\ \beta \in \mathbb{R},\ \forall v_1, v_2 \in H$,

$$\Phi(\alpha v_1 + \beta v_2) = \alpha\Phi(v_1) + \beta\Phi(v_2).$$

Φ is a bounded (i.e. continuous) functional. Indeed, $\forall v \in H$,

$$|\Phi(v)| \leq \|f\|_{L^2(G)}\|v\|_{L^2(G)}$$

due to the CBS inequality, so

$$|\Phi(v)| \leq \|f\|_{L^2(G)}C_{PF}\|\nabla v\|_{L^2(G)}, \qquad (2.4.6)$$

$$|\Phi(v)| \leq \|f\|_{L^2(G)}C_{PF}\frac{1}{\sqrt{\kappa}}[|v|] \qquad (2.4.7)$$

(we applied the PF inequality and then **(ii)**).

3. So, due to the Riesz–Frechet theorem, there exists a unique $u \in H$ such that

$$\forall v \in H,\quad \Phi(v) = [u, v],$$

i.e. $\forall v \in H_0^1(G)$, (2.4.3) holds, and the existence and uniqueness of a weak solution is proved.

4. Taking $v = u$, we get

$$[u, u] = \Phi(u) \leq C_{PF}\frac{1}{\sqrt{\kappa}}\|f\|_{L^2(G)}\sqrt{[u, u]}$$

(see (2.4.7)). So,

$$[|u|] \leq \frac{C_{PF}}{\sqrt{\kappa}}\|f\|_{L^2(G)}.$$

On the other side,

$$[|u|] = \sqrt{[u, u]} \geq \sqrt{\kappa}\sqrt{\|\nabla u\|_{L^2(G)}^2} = \sqrt{\kappa}\|\nabla u\|_{L^2(G)}$$

$$\geq \sqrt{\kappa}\frac{1}{C_{PF}}\|u\|_{L^2(G)},$$

and so

$$[|u|] \geq \frac{\sqrt{\kappa}}{\sqrt{2}}\min\left\{1, \frac{1}{C_{PF}}\right\}\|u\|_{H^1(G)}.$$

So, the theorem is proved:

$$\|u\|_{H^1(G)} \leq \frac{\sqrt{2}}{\kappa} C_{PF} \max\{C_{PF}, 1\} \|f\|_{L^2(G)}.$$

$\square$

Remark 4.1. Consider formally equation (2.4.1) with the right-hand side

$$f(x) = f_0(x) - \sum_{i=1}^{d} \frac{\partial f_i}{\partial x_i}(x), \qquad (2.4.8)$$

where $f_0, f_i (i = 1, \ldots, d)$ belong to $L^2(G)$. Let us define a weak solution of the problem (2.4.1), (2.4.2) as a function $u \in H_0^1(G)$ such that $\forall v \in C_0^{(\infty)}$,

$$\int_G A(x) \nabla u \cdot \nabla v \, dx = \int_G f_0(x) v(x) + \sum_{i=1}^{d} f_i(x) \frac{\partial v}{\partial x_i}(x) \, dx. \qquad (2.4.9)$$

Then, Theorem 4.1 remains valid for this case, and (2.4.4) is modified as

$$\|u\|_{H^1(G)} \leq C_D' \left\{ C_{PF} \|f_0\|_{L^2(G)} + \sqrt{\sum_{i=1}^{d} \|f_i\|_{L^2(G)}^2} \right\},$$

$$C_D' = \frac{\sqrt{2}}{\kappa} \max\{1, C_{PF}\}. \qquad (2.4.10)$$

Remark 4.2. In the case of regular right-hand side f and piecewise smooth coefficients A_{ij} having discontinuities at some smooth surfaces Σ, the weak solution satisfies equation (2.4.1) everywhere out of Σ and the interface conditions are

$$[u]_{\Sigma} = 0, \qquad \left[\sum_{i,j=1}^{d} A_{ij} \frac{\partial u}{\partial x_j} n_i \right]_{\Sigma} = 0$$

at Σ. If ∂G is smooth, then the weak solution satisfies the boundary condition (2.4.2), and so the weak solution coincides with the classical one. The theorems providing this assertion are called Agmon–Duglis–Nirenberg (ADN) theorems.

5. Stationary Elasticity Equation

Formally, the elasticity equation is similar to the conductivity equation written in the form $(2.4.1')$:

$$-\sum_{i,j=1}^{d} \frac{\partial}{\partial x_i}\left(A_{ij}(x)\frac{\partial \mathbf{u}}{\partial x_j}\right) = \mathbf{f}(x), \quad x \in G; \qquad (2.5.1)$$

however, $\mathbf{u}$ and $\mathbf{f}$ are vector-valued functions and every $A_{ij}(x)$ is a $d \times d$ matrix having entries $a_{ij}^{kl}(x)$. These entries are measurable bounded functions satisfying:

(i) $\forall x \in G, \ \ \forall i,j,k,l \in \{1,\ldots,d\}, \ \ a_{ij}^{kl}(x) = a_{ji}^{lk}(x) = a_{kj}^{il}(x);$
(ii) $\exists \kappa > 0 : \forall x \in G, \ \ \forall \eta = (\eta_{jl})_{1 \le j,l \le d} \in \mathbb{R}_{\text{sym}}^{d \times d},$

$$\sum_{i,j,k,l=1}^{d} a_{ij}^{kl}(x)\eta_{jl}\eta_{ik} \ge \kappa \sum_{j,l=1}^{d}(\eta_{jl})^2.$$

Here, $\mathbb{R}_{\text{sym}}^{d \times d}$ is the space of symmetric $d \times d$ matrices.

Consider the Dirichlet boundary-value problem for $(2.5.1)$:

$$\mathbf{u}\big|_{\partial G} = \mathbf{0}. \qquad (2.5.2)$$

As in the previous section, suppose that G is a bounded domain, $\mathbf{f} \in L^2(G)$. Here, and in the following, a vector-valued function belongs to a space H if all its entries belong to this space.

A weak solution is defined as a function $\mathbf{u} \in H_0^1(G)$ such that $\forall \mathbf{v} \in C_0^{(\infty)}(G),$

$$\int_G \sum_{i,j=1}^{d} A_{ij}(x)\frac{\partial \mathbf{u}}{\partial x_j} \cdot \frac{\partial \mathbf{v}}{\partial x_j} dx = \int_G \mathbf{f}(x) \cdot \mathbf{v}(x)dx. \qquad (2.5.3)$$

Using the same approach as in the previous section, we can prove the following theorem.

Theorem 5.1. *There exists a unique solution of problem* $(2.5.1)$, $(2.5.2)$ *and it satisfies the a priori estimate*

$$\|\mathbf{u}\|_{H^1(G)} \le C_D^e \|\mathbf{f}\|_{L^2(G)}, \qquad (2.5.4)$$

where the constant C_D^e is defined by κ, C_{PF}.

The difference in the proof with respect to Theorem 4.1 concerns the proof of the symmetry of the inner product

$$[\mathbf{u}, \mathbf{v}] = \int_G \sum_{i,j=1}^d A_{ij}(x) \frac{\partial \mathbf{u}}{\partial x_j} \cdot \frac{\partial \mathbf{v}}{\partial x_i} dx,$$

and the property $[\mathbf{v}, \mathbf{v}] = 0 \Leftrightarrow \mathbf{v} = \mathbf{0}$.

Indeed,

$$[\mathbf{u}, \mathbf{v}] = \int_G \sum_{i,j,k,l=1}^d a_{ij}^{kl}(x) e_{jl}(\mathbf{u}) e_{ik}(\mathbf{v}) dx, \qquad (2.5.5)$$

where

$$e_{jl}(\mathbf{u}) = \frac{1}{2} \left(\frac{\partial u_j}{\partial x_l} + \frac{\partial u_l}{\partial x_j} \right),$$

and taking in consideration **(i)**, we get

$$[\mathbf{u}, \mathbf{v}] = [\mathbf{v}, \mathbf{u}].$$

For the property $[\mathbf{v}, \mathbf{v}] = 0 \Leftrightarrow \mathbf{v} = \mathbf{0}$, we need *Korn's inequality*.

Lemma 5.1. (*Korn's inequality*) $\forall \mathbf{u} \in H_0^1(G)$,

$$E_G(\mathbf{u}) = \int_G \sum_{j,l=1}^d (e_{jl}(\mathbf{u}))^2 \, dx \geq \frac{1}{2} \|\nabla u\|_{L^2(G)}^2. \qquad (2.5.6)$$

Proof. Let G be a parallelepiped, and let $\mathbf{u}$ be $C_0^{(\infty)}$-smooth (belongs to $C_0^{(\infty)}(G)$). Then,

$$\int_G \frac{\partial u_i}{\partial x_j} \frac{\partial u_j}{\partial x_i} dx = - \int_G \frac{\partial^2 u_i}{\partial x_i \partial x_j} u_j dx = \int_G \frac{\partial u_i}{\partial x_i} \frac{\partial u_j}{\partial x_j} dx$$

for $i \neq j$.

Then,

$$E_G(\mathbf{u}) = \frac{1}{4}\int_G \sum_{i,j=1}^d \left\{ \left(\frac{\partial u_i}{\partial x_j}\right)^2 + 2\frac{\partial u_i}{\partial x_j}\frac{\partial u_j}{\partial x_i} + \left(\frac{\partial u_j}{\partial x_i}\right)^2 \right\} dx$$

$$= \frac{1}{4}\int_G \sum_{i,j=1}^d \left\{ \left(\frac{\partial u_i}{\partial x_j}\right)^2 + 2\frac{\partial u_i}{\partial x_i}\frac{\partial u_j}{\partial x_j} + \left(\frac{\partial u_j}{\partial x_i}\right)^2 \right\} dx$$

$$\geq \frac{1}{2}\int_G \sum_{i,j=1}^d \left(\frac{\partial u_i}{\partial x_j}\right)^2 + (\operatorname{div} \mathbf{u})^2 dx$$

$$\geq \frac{1}{2}\|\nabla u\|_{L^2(G)}^2.$$

Let now G be an arbitrary domain and $\mathbf{u} \in H_0^1(G)$. Then, extending it by zero to some parallelepiped and approximating by a sequence of $C_0^{(\infty)}$-smooth functions $\mathbf{u}_n$ (as in the proof of Theorem 3.1), we pass to the limit and prove estimate (2.5.6).
The lemma is proved. $\qquad\qquad\qquad\qquad\qquad\qquad\qquad\qquad$ □

This lemma allows us to obtain

$$[\mathbf{u}, \mathbf{u}] \geq \kappa E_G(\mathbf{u}) \geq \frac{\kappa}{2}\|\nabla u\|_{L^2(G)}^2.$$

Following the proof of Theorem 4.1, we get for C_D^e the formula $C_D^e = 2C_D$.

6. Stationary Stokes Equation

Consider a bounded domain G and the Dirichlet boundary value problem for Stokes equation:

$$-\nu \triangle \mathbf{v} + \nabla p = \mathbf{f}(x), \quad x \in G, \qquad\qquad (2.6.1)$$

$$\operatorname{div} \mathbf{v} = 0, \quad x \in G, \qquad\qquad (2.6.2)$$

$$\mathbf{v}\big|_{\partial G} = \mathbf{0}. \qquad\qquad (2.6.3)$$

Let $\mathbf{f}$ be a vector-valued function of $L^2(G)$, $\nu > 0$.
The weak formulation can be given in two different forms: "without pressure" or "with pressure".

The first formulation uses the subspace H_{div} of the Sobolev space $H_0^1(G)$ such that its functions are divergence free:

$$H_{\mathrm{div}} = \{\mathbf{v} \in H_0^1(G) \big| \operatorname{div} \mathbf{v} = 0\}.$$

Then, multiplying equation (2.6.1) by a test function $\mathbf{w} \in H_{\mathrm{div}}$ and integrating by parts over G, we get the following definition of a weak solution: $\mathbf{v} \in H_{\mathrm{div}}$, satisfying for all $\mathbf{w} \in H_{\mathrm{div}}$ the following integral identity:

$$\int_G \nu \nabla \mathbf{v} \cdot \nabla \mathbf{w} dx = \int_G \mathbf{f}(x) \cdot \mathbf{w}(x) dx. \qquad (2.6.4)$$

Repeating the proof of Theorem 4.1, we prove the following theorem. □

Theorem 6.1. *There exists a unique solution of* (2.6.4) *and*

$$\|\mathbf{v}\|_{H^1(G)} \leq C_D \|\mathbf{f}\|_{L^2(G)}.$$

In (2.4.5), κ *should be replaced by* ν.

Another definition of a weak solution "with pressure" is a couple $(\mathbf{v}, p)$, $\mathbf{v} \in H_{\mathrm{div}}$, $p \in L^2(G)$ such that for all $\mathbf{w} \in H_0^1(G)$,

$$\int_G \nu \nabla \mathbf{v} \cdot \nabla \mathbf{w} dx - \int_G p \operatorname{div} \mathbf{w} dx = \int_G \mathbf{f}(x) \cdot \mathbf{w}(x) dx. \qquad (2.6.5)$$

This definition is equivalent to the previous one in the sense that $\mathbf{v}$ is the same and one can prove that there exists a unique pressure p (unique up to an additive constant), satisfying (2.6.5).

The Navier–Stokes equation for small data also admits a unique solution, but the proof of the corresponding theorem is based on a fixed point theorem [2].

7.   Galerkin Method for the Heat Equation

Consider the following problem for the heat equation:

$$C(x)\frac{\partial u}{\partial t} - \text{div } (A(x)\nabla u) = f(x,t), \quad x \in G,\ t > 0, \quad (2.7.1)$$

$$u\big|_{\partial G} = 0, \quad (2.7.2)$$

$$u\big|_{t=0} = 0, \quad (2.7.3)$$

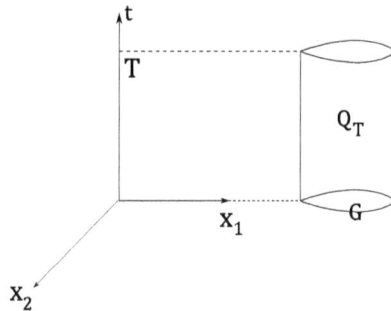

where G, A are the same as in Section 4, $f \in L^2(Q_T)$, $Q_T = G \times (0,T)$, and C is a bounded measurable function such that

(iii) $\exists \kappa > 0 : \forall x \in G,\ C(x) \geq \kappa.$

Let us prove that there exists a unique weak solution of (2.7.1)–(2.7.3).

Let us say that a function f of $C^{(\infty)}(\bar{Q}_T)$ vanishes in the neighborhood of $\partial G \times [0,T]$ if there exists $\varepsilon > 0$ such that $f(x,t) = 0$ $\forall x \in G_\varepsilon,\ \forall t \in [0,T]$. Denote $C_{G,0}^{(\infty)}(\bar{Q}_T)$ as the space of functions vanishing in the neighborhood of $\partial G \times [0,T]$, and let $H_{G,0}^1(Q_T)$ be its completion in $H^1(Q_T)$ norm

$$\sqrt{\|u\|_{L^2(Q_T)}^2 + \left\|\frac{\partial u}{\partial t}\right\|_{L^2(Q_T)}^2 + \sum_{i=1}^{d}\left\|\frac{\partial u}{\partial x_i}\right\|_{L^2(Q_T)}^2}.$$

Define as a weak solution of (2.7.1)–(2.7.3) the function $u \in H_{G,0}^1(Q_T)$ such that $u|_{t=0} = 0$ and the integral identity holds:

$\forall v \in H_0^1(G)$,

$$\int_G C(x)\frac{\partial u}{\partial t}v(x) + A(x)\nabla u \cdot \nabla v dx = \int_G f(x,t)v(x)dx. \qquad (2.7.4)$$

Here, ∇ is the gradient with respect to x.

Consider a base $\mathcal{B} = (\psi_1, \psi_2, \ldots)$ in $H_0^1(G)$. Then, every function of $H_0^1(G)$ can be uniquely presented in the form of the sum

$$\sum_{i=1}^{\infty} \alpha_i \psi_i, \quad \alpha_i \in \mathbb{R}.$$

The *Galerkin method* approximates the problem (2.7.4) by its projection on the subspace H_N of $H_0^1(G)$ which is the set of linear combinations of $\psi_1, \ldots, \psi_N$:

$$H_N = \text{span}\,(\psi_1, \ldots, \psi_N).$$

Namely, we seek a solution of an approximated Galerkin problem in the form

$$u_N(x,t) = \sum_{i=1}^{N} \alpha_i(t)\psi_i(x), \qquad (2.7.5)$$

where $\alpha_i \in H^1((0,T))$.

Taking in (2.7.4) test functions $v = \psi_j$, we get

$$\sum_{i=1}^{N}\left(\int_G C(x)\psi_i\psi_j dx\right)\dot{\alpha}_i(t) + \sum_{i=1}^{N}\left(\int_G A(x)\nabla\psi_i \cdot \nabla\psi_j dx\right)\alpha_i(t)$$

$$= \int_G f(x,t)\psi_j(x)dx, \quad j = 1, \ldots, N. \qquad (2.7.6)$$

This is a linear system of ordinary-differential equations with constant coefficients. It is well known that if we add the initial condition

$$\alpha_i(0) = 0, \qquad (2.7.7)$$

then this system admits a unique solution

$$(\alpha_1, \ldots, \alpha_N) \in \left(H^1((0,T))\right)^N.$$

One can prove that the sequence of Galerkin's approximations $(u_N)_{N \in \mathbb{N}}$ converges to the solution u of problem (2.7.4).

Let us check that it weakly converges to u.

(1) Multiplying equations (2.7.6) by α_j and summing up, we obtain

$$\int_G C(x)\frac{\partial u_N}{\partial t}u_N dx + \int_G A(x)\nabla u_N \cdot \nabla u_N dx = \int_G f u_N dx. \quad (2.7.8)$$

Integrating (2.7.8) in time over $[0,T]$, we get

$$\frac{1}{2}\int_G C(x)(u_N)^2|_{t=T}dx + \int_0^T \int_G A(x)\nabla u_N \cdot \nabla u_N dxdt$$

$$= \int_0^T \int_G f u_N dxdt \leq \|f\|_{L^2(Q_T)}\|u_N\|_{L^2(Q_T)}$$

$$\leq C_{PF}\|f\|_{L^2(Q_T)}\|\nabla u_N\|_{L^2(Q_T)}. \quad (2.7.9)$$

Taking into account that the first integral is non-negative and the second integral in the left-hand side is greater than $\kappa\|\nabla u_N\|_{L^2(Q_T)}^2$, we get

$$\|\nabla u_N\|_{L^2(Q_T)} \leq \frac{C_{PF}}{\kappa}\|f\|_{L^2(Q_T)}. \quad (2.7.10)$$

(2) Multiplying equations (2.7.6) by $\dot{\alpha}_j$, summing up and integrating in time over $[0,T]$, we get

$$\int_0^T \int_G C(x)\left(\frac{\partial u_N}{\partial t}\right)^2 dxdt + \frac{1}{2}\int_G A(x)\nabla u_{N|t=T} \cdot \nabla u_{N|t=T}dx$$

$$= \int_0^T \int_G f(x,t)\frac{\partial u_N}{\partial t}dxdt \leq \|f\|_{L^2(Q_T)}\left\|\frac{\partial u_N}{\partial t}\right\|_{L^2(Q_T)}, \quad (2.7.11)$$

and so

$$\left\|\frac{\partial u_N}{\partial t}\right\|_{L^2(Q_T)} \leq \frac{1}{\kappa}\|f\|_{L^2(Q_T)}. \quad (2.7.12)$$

Taking into account that

$$\|u_N\|_{L^2(Q_T)} \leq C_{PF}\|\nabla u_N\|_{L^2(Q_T)},$$

we see that $(u_N)_{N\in\mathbb{N}}$ is bounded in $H^1(Q_T)$, and so it contains a subsequence weakly convergent to some function $u \in H^1(Q_T)$ (due to the Banach–Alaoglu theorem; see Theorem 1.7). The function u satisfies (2.7.4).

Indeed, it can be proved that the space of linear combinations (span)

$$\mathcal{M} = \left\{ \sum_{i=1}^{M} \beta_i(t)\psi_i(x), \quad \beta_i \in H^1((0,T)) \right\}$$

is dense in $H^1_{G,0}(Q_T)$.

Let v be a function of $\mathcal{M}$ equal to

$$v(x,t) = \sum_{i=1}^{M} \beta_i(t)\psi_i(x).$$

Multiply equations (2.7.6) by $\beta_j(t)$ and integrate in time over $(0,T)$. Then, passing to the limit in $N \to +\infty$, we see that the weak limit u of a subsequence of $(u_N)_{N \in \mathbb{N}}$ satisfies the identity $\forall v \in \mathcal{M}$:

$$\int_0^T \int_G C(x)\frac{\partial u}{\partial t}v(x,t) + A(x)\nabla u \cdot \nabla v dx dt = \int_0^T \int_G f(x,t)v(x,t)dx dt,$$

and so ($\mathcal{M}$ is dense in $H^1_{G,0}(Q_T)$) it remains valid for all functions $v \in H^1_{G,0}(Q_T)$.

This identity yields (2.7.4) (for almost all $t \in [0,T]$).

The uniqueness can be proved using the same approach that we used for (2.7.10), (2.7.12) giving the estimate

$$\|u\|_{H^1(Q_T)} \leq \frac{\sqrt{1 + C_{PF}^2 + C_{PF}^4}}{\kappa}\|f\|_{L^2(Q_T)}. \tag{2.7.13}$$

Thus, all subsequences of $(u_N)_{N \in \mathbb{N}}$ converge weakly to u, so $(u_N)_{N \in \mathbb{N}}$ also converges weakly to u.

The Galerkin method is used not only for theoretical objectives (proof of the existence of solution) but also as a numerical method, called the *finite element method* (*FEM*). Various versions of this method differ by the choice of the base of H_N. The simplest version uses functions ψ_j which are so-called hat functions.

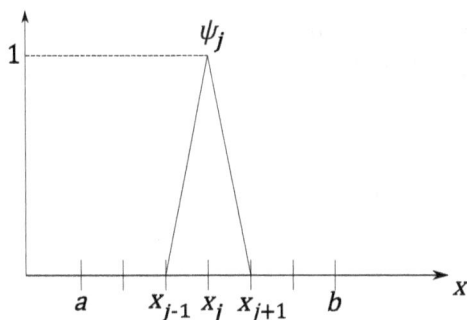

In the case of $d = 1$, the interval $G = (a, b)$ is divided into $N + 1$ small pieces: (x_i, x_{i+1}), where $x_i = a + ih$, $i = 0, \ldots, N$, $h = \frac{b-a}{N}$, and ψ_j are defined as follows:

$$\psi_j(x) = \frac{x - x_{j-1}}{h} \quad \text{on } [x_{j-1}, x_j],$$

$$\psi_j(x) = \frac{x_{j+1} - x}{h} \quad \text{on } [x_j, x_{j+1}],$$

and

$$\psi_j(x) = 0 \quad \text{out of } [x_{j-1}, x_{j+1}].$$

8.   On the Finite Difference Method

Generally, the partial differential equations are solved numerically by the finite element method, finite difference method, or finite volume methods. The finite element method is a version of the Galerkin method. Let us consider the *finite difference method* in the one-dimensional case, where G is an interval $(0, 1)$.

8.1.   *Approximation of the heat equation by an explicit finite difference scheme*

Consider the boundary-value problem for the heat equation

$$C\frac{\partial u}{\partial t} - K\frac{\partial^2 u}{\partial x^2} = f(x, t), \quad (x, t) \in (0, 1) \times (0, +\infty), \quad (2.8.1)$$

$$u(0, t) = 0, \quad u(1, t) = 0, \quad (2.8.2)$$

$$u(x, 0) = \psi(x), \quad (2.8.3)$$

where $C, K > 0, f, \psi$ are smooth functions (such that $\frac{\partial^2 u}{\partial t^2}$ and $\frac{\partial^4 u}{\partial x^4}$ are continuous in $G = [0,1] \times [0,+\infty)$).

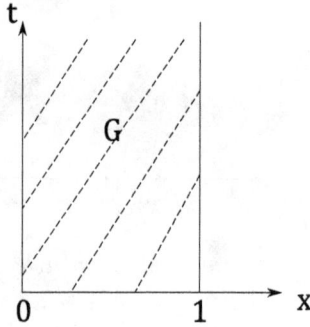

Approximate G by a mesh:

$$G_{h,\tau} = \{(x_m, t^n) = (mh, \tau n), \quad m \in \{0, \dots, M\}, \quad n \in \mathbb{N}\}, \quad M = 1/h.$$

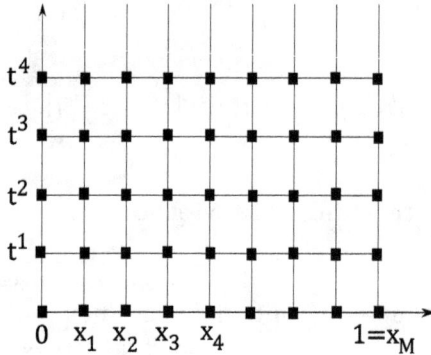

Approximate

$$\left.\frac{\partial u}{\partial t}\right|_{(x_m,t^n)} \approx \frac{u(x_m, t^n + \tau) - u(x_m, t^n)}{\tau},$$

$$\left.\frac{\partial^2 u}{\partial x^2}\right|_{(x_m,t^n)} \approx \frac{u(x_{m+1}, t^n) - 2u(x_m, t^n) + u(x_{m-1}, t^n)}{h^2}.$$

Order of approximation is $O(\tau + h^2)$, more exactly,

$$Cu_t - Ku_{\bar{x}x} = f(x_m, t^n) + r_{m,n}, \quad 1 \leq m \leq M - 1, \quad t^n \leq T,$$

$$|r_{m,n}| \leq \left(C\tau + \frac{Kh^2}{12}\right) \bar{U}_T, \tag{2.8.4}$$

where

$$\bar{U}_T = \max \left\{ \sup_{(x,t)\in[0,1]\times[0,T]} \left| \frac{\partial^2 u}{\partial t^2}(x,t) \right|, \right.$$
$$\left. \sup_{(x,t)\in[0,1]\times[0,T]} \left| \frac{\partial^4 u}{\partial x^4}(x,t) \right| \right\}. \qquad (2.8.5)$$

Replace (2.8.1)–(2.8.3) by a finite difference scheme (FDS):

$$C\frac{y_m^{n+1} - y_m^n}{\tau} - K\frac{y_{m+1}^n - 2y_m^n + y_{m-1}^n}{h^2} = f_m^n, \qquad (2.8.6)$$
$$1 \leq m \leq M - 1, \quad 0 \leq n \leq N - 1,$$
$$y_0^{n+1} = 0, \quad y_M^{n+1} = 0, \quad n = 0, \ldots, N - 1, \qquad (2.8.7)$$
$$y_m^0 = \psi_m, \quad m = 0, \ldots, M, \qquad (2.8.8)$$

where $f_m^n = f(x_m, t^n)$, $\psi_m = \psi(x_m)$ ($\psi_0 = \psi_M = 0$), $N = T/\tau$ (T is a finite number).

Algorithm of the solution:

```
for (int m = 0;  M;  step  1)
{y_m^0 = ψ_m; };
for (int n = 0;  N − 1;  step  1)

{for (int m = 1;  M − 1;  step 1)
    {y_m^1 = y_m^0 + (τ/C)K(y_{m+1}^0 − 2y_m^0 + y_{m−1}^0)/h² + (τ/C)f_m^n
    };
    y_0^1 = 0  y_M^1 = 0;
    for (int m = 0;  M;  step 1)
    {y_m^0 = y_m^1};
};
```

8.1.1. Stability of the difference scheme

Consider the spaces:

$E_M = \mathbb{R}^{M-1}$ supplied with the inner product

$$(y, z)_{l^2} = h \sum_{j=1}^{M-1} y_j z_j, \quad y = (y_1, \ldots, y_{M-1}), \quad z = (z_1, \ldots, z_{M-1}),$$

$$E_{M,N} = \{y^0, \ldots, y^N \mid y^n \in E_M, n = 0, \ldots, N\},$$

supplied with the norm

$$\|y\|_{l^{1,2}} = \sum_{n=0}^{N} \tau \|y^n\|_{l^2}.$$

Let $N\tau = T$ be a finite real number, and let the following *Courant–Friedrichs–Lewy (CFL) condition* hold:

$$\frac{\tau K}{h^2 C} \leq \frac{1}{2}.$$

Let us prove the stability of the difference scheme (2.8.6)–(2.8.8): there exists a positive number C independent of h, τ such that for sufficiently small h, τ, the estimate holds:

$$\max_{0 \leq n \leq N} \|y^n\|_{l^2} \leq C \left(\|\psi\|_{l^2} + \|f\|_{l^{1,2}} \right). \tag{2.8.9}$$

Consider the following $(M-1) \times (M-1)$ matrix:

$$\Lambda_h = \frac{1}{h^2} \begin{pmatrix} 2 & -1 & 0 & \ldots & 0 & 0 \\ -1 & 2 & -1 & \ldots & 0 & 0 \\ .. & .. & .. & \ldots & .. & .. \\ 0 & 0 & 0 & \ldots & -1 & 2 \end{pmatrix},$$

and note that it is positive definite having $N-1$ positive eigenvalues $\{\lambda_1, \ldots, \lambda_{M-1}\}$,

$$\lambda_k = \frac{4}{h^2} \sin^2 \left(\frac{k\pi h}{2} \right),$$

$0 < \lambda_1 < \lambda_2 < \cdots < \lambda_{M-1} < 4/h^2$, associated to eigenvectors

$$\mu^{(t)} = \left(\mu_1^{(k)}, \ldots, \mu_{M-1}^{(k)}\right), \quad \mu_j^{(k)} = \sqrt{2}\sin(\pi k j h), \quad j = 1, \ldots, M-1.$$

Indeed, the jth entry of the vector $\Lambda_h \mu^{(k)}$ is equal to

$$\left(\Lambda_h \mu^{(k)}\right)_j = -\frac{\mu_{j+1}^{(k)} - 2\mu_j^{(k)} + \mu_{j-1}^{(k)}}{h^2}$$

$$= -\sqrt{2}\frac{\sin(\pi k(j+1)h) - 2\sin(\pi k j h) + \sin(\pi k(j-1)h)}{h^2}$$

$$= \sqrt{2}\left(\frac{2 - 2\cos(k\pi h)}{h^2}\right)\sin(\pi k j h) = \frac{4}{h^2}\sin^2\left(\frac{k\pi h}{2}\right)\mu_j^{(k)}.$$

Here, and in the following, the entry numbers 0 and M are equal to 0. $\mu^{(k)}$ are orthonormal with respect to $(\cdot, \cdot)_{l^2}$.

Let us expand ψ, $f(\cdot, t^n)$, and y^n on the grid with respect to the orthonormal base $(\mu^{(1)}, \ldots, \mu^{(M-1)})$:

$$\psi = \sum_{k=1}^{M-1} \alpha_k \mu^{(k)}, \quad f^n = \sum_{k=1}^{M-1} \beta_k^{(n)} \mu^{(k)}, \quad y^n = \sum_{k=1}^{M-1} q_k^{(n)} \mu^{(k)},$$

and note that

$$\sum_{k=1}^{M-1} \alpha_k^2 = \|\psi\|_{l^2}^2, \quad \sum_{k=1}^{M-1} \left(\beta_k^{(n)}\right)^2 = \|f^n\|_{l^2}, \quad \text{and} \quad \|y^n\|_{l^2}^2 = \sum_{k=1}^{M-1} \left(q_k^{(n)}\right)^2.$$

The difference scheme (2.8.6) has the form

$$\sum_{k=1}^{M-1}\left(\frac{C}{\tau}\left(q_k^{(n+1)} - q_k^{(n)}\right) + K q_k^{(n)}\Lambda_h\right)\mu^{(k)} = \sum_{k=1}^{M-1} \beta_k^{(n)}\mu^{(k)},$$

i.e.

$$\sum_{k=1}^{M-1}\left(\frac{C}{\tau}\left(q_k^{(n+1)} - q_k^{(n)}\right) + \lambda_k K q_k^{(n)}\right)\mu^{(k)} = \sum_{k=1}^{M-1} \beta_k^{(n)}\mu^{(k)},$$

i.e.

$$q_k^{(n+1)} = \left(1 - \frac{K\tau}{C}\lambda_k\right)q_k^{(n)} + \frac{\tau}{C}\beta_k^{(n)},$$

i.e.

$$\sum_{k=1}^{M-1} q_k^{(n+1)} \mu^{(k)} = \sum_{k=1}^{M-1} \left(1 - \frac{K\tau}{C}\lambda_k\right) q_k^{(n)} \mu^{(k)} + \sum_{k=1}^{M-1} \frac{\tau}{C}\beta_k^{(n)} \mu^{(k)},$$

$$\|y^{(n+1)}\|_{l^2} \leq \left\|\sum_{k=1}^{M-1} \left(1 - \frac{K\tau}{C}\lambda_k\right) q_k^{(n)} \mu^{(k)}\right\|_{l^2} + \frac{\tau}{C}\|f^n\|_{l^2}$$

$$\leq \max_{1 \leq k \leq M-1} \left|1 - \frac{K\tau}{C}\lambda_k\right| \|y^{(n)}\|_{l^2} + \frac{\tau}{C}\|f^n\|_{l^2}.$$

Assume that the CFL condition is satisfied. Then,

$$1 - \frac{4K\tau}{Ch^2} \leq 1 - \frac{K\tau}{C}\lambda_k \leq 1,$$

and so

$$-1 \leq 1 - \frac{K\tau}{C}\lambda_k \leq 1,$$

i.e.

$$\left|1 - \frac{K\tau}{C}\lambda_k\right| \leq 1.$$

So,

$$\|y^{(n+1)}\|_{l^2} \leq \|y^n\|_{l^2} + \frac{\tau}{C}\|f^n\|_{l^2}$$

and

$$\max_{0 \leq n \leq N} \|y^n\|_{l^2} \leq \|\psi\|_{l^2} + \sum_{n=0}^{N-1} \frac{\tau}{C}\|f^n\|_{l^2},$$

$$\max_{0 \leq n \leq N} \|y^n\|_{l^2} \leq \|\psi\|_{l^2} + C^{-1}\|f\|_{l^{1,2}}.$$

It can be proved that the stability remains true if CFL is replaced by

$$\frac{\tau K}{h^2 C} \leq \frac{1}{2} + C_1\tau,$$

where C_1 is independent of τ, h.

Stability + Consistency ⇒ Convergence.

Consider $w_m^n = u_m^n - y_m^n$. It satisfies FDS

$$
\begin{cases}
C\frac{w_m^{n+1}-w_m^n}{\tau} - K\frac{w_{m+1}^n-2w_m^n+w_{m-1}^n}{h^2} = r_{m,n}, \\
1 \le m \le M-1, \quad 0 \le n \le N-1, \\
w_0^{n+1} = 0, \quad w_M^{n+1} = 0 \quad n = 0,\dots,N-1, \\
w_m^0 = 0, \quad m = 0,\dots,M,
\end{cases}
$$

and so due to stability, we get

$$
\max_{0\le n\le N} \|w^n\|_{l^2} \le C^{-1}\|r\|_{l^{1,2}}.
$$

Due to consistency (2.8.4), we get now the convergence

$$
\max_{0\le n\le N} \|u^n - y^n\|_{l^2} = O(\tau + h^2).
$$

8.2. *Generalizations: Formal spectral rule of stability (linear case)*

In the case where a difference scheme has a solution y_m^n, we consider the right-hand side equal to zero. Plugging in the scheme a particular solution of the form

$$
y_m^n = e^{i\alpha m} \cdot q^n \quad (n \text{ is the power}),
$$

we factorize this expression and get an equation for $q \in \mathbb{C}$ with coefficients depending on τ, h, and $\alpha \in \mathbb{R}$.

The *spectral rule of stability*: the difference scheme is stable if for all $\alpha \in \mathbb{R}$ all roots of the equation for q satisfy the inequality

$$
|q| \le 1 \quad (\text{SSR} - \text{strict spectral rule}),
$$

or

$$
|q| \le 1 + C_1\tau \quad (\text{SR} - \text{weakened form}).
$$

Here, C_1 does not depend on h, τ.

Example 1. Explicit scheme for the heat equation:

$$C\frac{y_m^{n+1} - y_m^n}{\tau} - K\frac{y_{m+1}^n - 2y_m^n + y_{m-1}^n}{h^2} = 0,$$

so

$$C\frac{e^{im\alpha}(q^{n+1} - q^n)}{\tau} - Kq^n\frac{e^{i(m+1)\alpha} - 2e^{im\alpha} + e^{i(m-1)\alpha}}{h^2} = 0,$$

i.e.

$$e^{im\alpha}q^n\left\{\frac{C(q-1)}{\tau} - K\frac{e^{i\alpha} - 2 + e^{-i\alpha}}{h^2}\right\} = 0,$$

i.e.

$$\frac{C(q-1)}{\tau} - K\frac{2\cos\alpha - 2}{h^2} = 0,$$

i.e.

$$q = 1 - \frac{4K\tau}{Ch^2}\sin^2\frac{\alpha}{2}.$$

Then,

$$\forall \alpha \in \mathbb{R}, \ |q| \leq 1 \Leftrightarrow -1 \leq 1 - \frac{4K\tau}{Ch^2} \Leftrightarrow \frac{K\tau}{Ch^2} \leq \frac{1}{2} \ \text{(CFL)}.$$

Example 2. Implicit scheme for the heat equation:

$$C\frac{y_m^{n+1} - y_m^n}{\tau} - K\frac{y_{m+1}^{n+1} - 2y_m^{n+1} + y_{m-1}^{n+1}}{h^2} = 0.$$

So,

$$C\frac{e^{im\alpha}(q^{n+1} - q^n)}{\tau} - Kq^{n+1}\frac{e^{i(m+1)\alpha} - 2e^{im\alpha} + e^{i(m-1)\alpha}}{h^2} = 0,$$

i.e.

$$e^{im\alpha}q^n\left\{\frac{C(q-1)}{\tau} - Kq\frac{e^{i\alpha} - 2 + e^{-i\alpha}}{h^2}\right\} = 0,$$

i.e.

$$\left(\frac{C}{\tau} + K\frac{4\sin^2\frac{\alpha}{2}}{h^2}\right)q = \frac{C}{\tau}.$$

So,

$$q = \frac{1}{1 + \frac{4\sin^2\frac{\alpha}{2}}{Ch^2}K\tau} \leq 1.$$

The scheme is unconditionally stable.

Example 3. Explicit scheme for the wave equation:

$$C\frac{y_m^{n+1} - 2y_m^n + y_m^{n-1}}{\tau^2} - K\frac{y_{m+1}^n - 2y_m^n + y_{m-1}^n}{h^2} = 0$$

(for the wave equation $C\frac{\partial^2 u}{\partial t^2} - K\frac{\partial^2 u}{\partial x^2} = 0$).

Consider $y_m^n = e^{i\alpha m}q^n$. We get

$$q^n e^{i\alpha m}\left(C\frac{q - 2 + q^{-1}}{\tau^2} - K\frac{e^{i\alpha} - 2 + e^{-i\alpha}}{h^2}\right) = 0,$$

i.e.

$$q^2 - 2\left(\left(1 - \frac{K}{C}\left(\frac{\tau}{h}\right)^2(1 - \cos\alpha)\right)\right)q + 1 = 0.$$

By the Vietta theorem, this equation has two solutions q_1, q_2 such that

$$q_1 q_2 = 1.$$

Let us apply the strict spectral rule ($|q_1| \le 1$ and $|q_2| \le 1$). So, $|q_1| = |q_2| = 1$.

Thus, $q_1 = e^{i\beta}$, $q_2 = e^{-i\beta}$, where $\beta \in \mathbb{R}$ is the new unknown. So,

$$C\frac{e^{i\beta} - 2 + e^{-i\beta}}{\tau^2} - K\frac{e^{i\alpha} - 2 + e^{-i\alpha}}{h^2} = 0,$$

i.e.

$$C\frac{\cos\beta - 1}{\tau^2} = K\frac{\cos\alpha - 1}{h^2},$$

i.e.

$$\cos\beta = 1 + \frac{K\tau^2}{Ch^2}(\cos\alpha - 1).$$

This equation must have a solution for all $\alpha \in \mathbb{R}$. So,

$$\max_{\alpha\in\mathbb{R}}\left|1 - \frac{2K\tau^2}{Ch^2}\sin^2\left(\frac{\alpha}{2}\right)\right| \le 1,$$

i.e.

$$-1 \le 1 - \frac{2K\tau^2}{Ch^2},$$

i.e.

$$\frac{K}{C}\left(\frac{\tau}{h}\right)^2 \le 1.$$

Example 4.

Consider the scheme

$$\frac{y_m^{n+1} - y_m^n}{\tau} + a\frac{y_m^n - y_{m-1}^n}{h} = 0, \quad a > 0$$

(for the transfer equation $\frac{\partial u}{\partial t} + a\frac{\partial u}{\partial x} = 0$). So,

$$e^{ima}\frac{q^{n+1} - q^n}{\tau} + aq^n\frac{e^{ima} - e^{i(m-1)\alpha}}{h} = 0,$$

$$q^n e^{ima}\left\{\frac{q-1}{\tau} + a\frac{1 - e^{-i\alpha}}{h}\right\} = 0,$$

$$q = 1 - \frac{a\tau}{h} + \frac{a\tau}{h}e^{-i\alpha}.$$

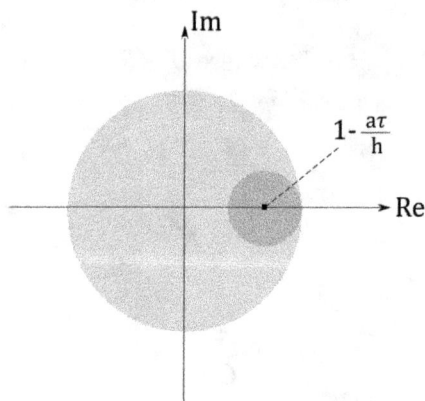

Stability condition:

$$1 - \frac{a\tau}{h} \geq 0,$$

i.e.

$$\frac{a\tau}{h} \leq 1.$$

References

[1] V.A. Trenogin. *The Functional Analysis*, Moscow: Nauka; 1980; French translation: V.A. Trenoguine. *Analyse fonctionnelle*, Éditions Mir, 1985.

[2] O.A. Ladyzhenskaya. *The Mathematical Theory of Viscous Incompressible Flow*, Moscow: Nauka; 1970; English edition: New York/ London/Paris: Gordon Breach Science Publishers, 1969.

Chapter 3

Homogenization: From Micro-scale to Macro-scale: Application to Mechanics of Composite Materials

1. What is a Composite Material?

We define here a *composite material* as a heterogeneous material constituting alternating volumes of some homogeneous materials at the "super-molecular scale." The shape classification of these materials can be presented by:

- stratified (laminated) media constituting (periodically) alternating homogeneous thin layers;
- fiber-reinforced materials: a periodic system of unidirectional parallel fibers of one compound (or a union of some systems of different directions) separated by a background material called matrix which fills the space between the fibers;
- granular composite: the periodic structure of a three-dimensional (3-D) periodic system of grains and a material which fills the space between the grains (the matrix).

This list can be continued.

It is assumed throughout that the dimensions of the periodic cell are much smaller than the characteristic macroscopic spacial size of the problem. If we take this macroscopic size as unity in the space dimension, then the period of the structure becomes a dimensionless, small parameter denoted as ε.

Composite materials have been widely applied throughout the 20th century in spacecraft, aircraft, and sports industries. These materials allowed to combine such properties as high stiffness, small density, high failure resistance, low heat conductivity, and radio transparency. The necessity of computation of macroscopic constants and properties starting from the microscopic shape information generated a lot of theories on the passing from the microscopic to the macroscopic scale (this procedure was called "averaging" and later *homogenization* or "up-scaling").

Apart from composite materials, there are many other examples of heterogeneous media (that can be also considered as a special case of the composite materials with vacuum as one compound): porous media, frames, grids, lattice structures, and atomic lattices. In particular, the modeling of flows in porous media is an important mathematical tool in oil and gas recovery engineering.

The standard numerical methods (such as the finite element method, finite difference method, and finite volume method) are inapplicable to these materials because all these methods use meshes which should be much finer than ε, and this leads to memory and time overflow for the majority of practical problems. The analytical methods are applicable mainly in the case of constant coefficients (which is not the case in composite materials) or at least in two-dimensional modeling (complex variable methods).

2.   From Micro to Macro

The first steps in the direction of a general theory of the micro to macro passage has been done by mechanicians at the beginning of the 20th century, although one can find some works of Maxwell related to this topic at the end of the 19th century. The main issue was to postulate that the heterogeneous at the microscopic scale material can be considered at the macroscopic scale as a homogeneous material. More precisely, the *equivalent homogeneity hypothesis* has been formulated which states that at the macroscopic scale, the heterogeneous material is mechanically (or physically) equivalent to some homogeneous one, and so the main problem is to calculate properties of this equivalent material.

For example, if a heterogeneous material consists of two different materials having the heat conductivity coefficients k_1 and k_2, then the main problem is to calculate the macroscopic conductivity $\hat{k}$ of the equivalent homogeneous material. The first attempts were related to the following means: arithmetic mean, harmonic mean or geometrical mean, i.e. if θ and $1-\theta$ are respectively the concentrations of the first and second compounds and $< f >$ denotes $f_1\theta + f_2(1-\theta)$, then the arithmetic mean of the conductivity is equal to $< k >$, the harmonic mean is equal to $< k^{-1} >^{-1}$, and the geometric mean is equal to $\exp\{< \ln(k) >\}$.

In particular, for $\theta = 0.5$, we get respectively $\frac{k_1+k_2}{2}$, $\frac{2}{\frac{1}{k_1}+\frac{1}{k_2}}$, and $\sqrt{k_1 k_2}$.

For each of these formulas, one can find the situation when it is correct and other situations when it is wrong. So, the idea was to identify a general approach and moreover to check the equivalent homogeneity hypothesis. Although this hypothesis has been assumed to be incontestable, we will give later some examples when it is wrong.

During the 1960s and 1970s of the 20th century, the first mathematical theories of up-scaling appeared. The main idea was to find an asymptotic solution for the partial derivative equation (PDE) describing a process, in which ε is a small positive parameter standing for the ratio l/L of the characteristic sizes of the micro- and macro-scales. This idea can also be the criterion for the applicability of the equivalent homogeneity hypothesis: this hypothesis is good if the solution for the PDE describing the heterogeneous medium is ε−close to a solution for some PDE with constant coefficients describing the homogeneous medium.

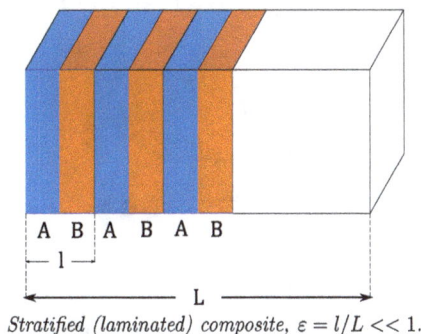

Stratified (laminated) composite, $\varepsilon = l/L \ll 1$.

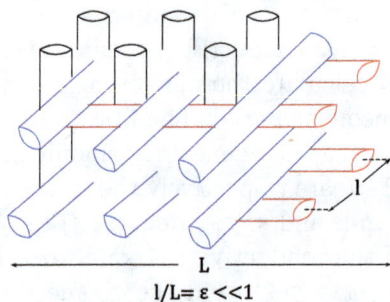

$$l/L = \varepsilon \ll 1$$

Fiber-reinforced composite.

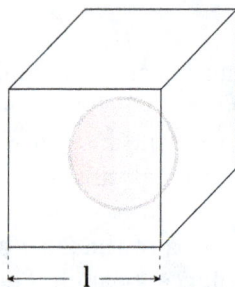

Periodic cell of a granular composite.

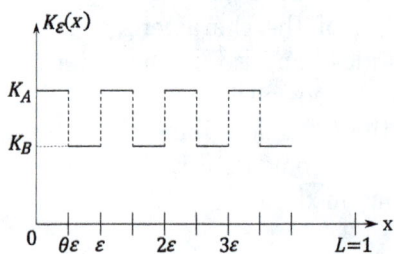

Conductivity as a function of x.

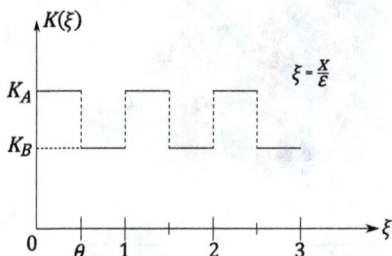

Conductivity as a function of ξ.

Consider a stratified material constituting thin alternating layers of materials A and B (and having conductivities K_A and K_B, respectively).

Let the macroscopic characteristic size be taken equal to 1, then the characteristic microscopic size l is equal to the small parameter ε.

Assume that θ is the concentration (volumetric rate) of the material A. Then, function $K_\varepsilon(x)$, the conductivity as a function of the coordinate x, has the following form:

$$K_\varepsilon(x) = \begin{cases} K_A \text{ if } x \in \bigcup_{i=0}^{N-1}[i\varepsilon, (i+\theta)\varepsilon), \\[2mm] K_B \text{ if } x \in \bigcup_{i=0}^{N-1}[(i+\theta)\varepsilon, (i+1)\varepsilon), \end{cases}$$

where $N = 1/\varepsilon$.

Let us pass to the dilated variable $\xi = x/\varepsilon$. One can see that

$$K_\varepsilon(x) = K\left(\frac{x}{\varepsilon}\right), \quad x \in [0, 1),$$

where

$$K(\xi) = \begin{cases} K_A \text{ if } \xi \in \bigcup_{i=-\infty}^{+\infty}[i, i+\theta), \\[2mm] K_B \text{ if } \xi \in \bigcup_{i=-\infty}^{+\infty}[i+\theta, i+1) \end{cases}$$

is a 1-periodic function of ξ,

$$\forall \xi \in \mathbb{R}, \quad K(\xi+1) = K(\xi).$$

Clearly, the same situation occurs in the 3-D case: the conductivity K_ε of the material as a function of $x = (x_1, x_2, x_3)$ is equal to

$$K_\varepsilon(x) = K\left(\frac{x}{\varepsilon}\right),$$

where $K(\xi)$ is a 1-periodic function of $\xi \in \mathbb{R}^3$, i.e. $\forall \xi \in \mathbb{R}^3$, $\forall j \in \mathbb{Z}^3$, $K(\xi+j) = K(\xi)$.

As an example, one can consider a PDE in mathematical physics describing the steady-state heat field in a composite material:

$$-div(K(\frac{x}{\varepsilon})\nabla u_\varepsilon) = f(x), \quad x = (x_1, x_2, x_3) \in \Omega \qquad (3.2.1)$$

set in a bounded domain $\Omega \subset \mathbb{R}^3$ having the characteristic size of order 1. Here, ε is a period of the micro-structure, and so $K(\xi)$ is a positive definite symmetric 1-periodic (3×3 matrix-valued) function of the dilated (fast, microscopic) variable ξ related to x by $\xi = x/\varepsilon$; it means that for all $\xi = (\xi_1, \xi_2, \xi_3) \in \mathbb{R}^3$, for all vectors $j = (j_1, j_2, j_3) \in \mathbb{Z}^3$ with integer components, the following relation holds: $K(\xi + j) = K(\xi)$. $K(\frac{x}{\varepsilon})$ stands for the conductivity matrix as a function of the space variable x; it takes different values depending on whether x belongs to the background material or inclusions; $u_\varepsilon(x)$ stands for an unknown temperature at the point x; domain Ω is assumed to be occupied by the composite material; f stands for the distribution of the heat sources.

Equation (3.2.1) should be equipped by a boundary condition (see, for example, Ref. [22]), say, of Dirichlet's type: $u_\varepsilon = 0$ on the boundary $\partial\Omega$. Then, the equivalent homogeneity hypothesis postulates the existence of a positive definite symmetric constant 3×3 matrix $\hat{K}$ such that, for any f that is smooth enough, solution u_0 of the so-called *homogenized equation* similar to (3.2.1) but with constant coefficient

$$-div(\hat{K}\nabla u_0) = f(x) \qquad (3.2.2)$$

equipped by the same boundary condition is close to u_ε in some norm

$$\|u_\varepsilon - u_0\| \leq \text{const } \varepsilon.$$

The homogenization method **proves (and not assumes)** the existence of this constant matrix $\hat{K}$ called the effective (macroscopic) conductivity; it gives an algorithm for computing this $\hat{K}$ and it gives the correctors to u_0 taking into account the fluctuations of the exact solution u_ε.

Remark 2.1. Continuity conditions are satisfied on the matrix-inclusion interface for temperature

$$[u_\varepsilon] = 0, \qquad (3.2.3)$$

and for the heat-flux density

$$\left[(K(\frac{x}{\varepsilon})\nabla u_\varepsilon, \mathbf{n}) \right] = 0, \qquad (3.2.4)$$

where the square brackets [ ] denote the jump of the function in transition through the interface and $\mathbf{n}$ stands for the normal vector to the interface.

3. Homogenization Techniques: Heat Equation (1-D case)

Consider a model one-dimensional (1-D) in space for a time-dependent problem. The homogenization method for this model case is described as follows. The heat equation for an unknown temperature $u_\varepsilon(x,t)$ depending on the space variable x and the time variable t is given by

$$L_\varepsilon u_\varepsilon \equiv c\left(\frac{x}{\varepsilon}\right)\frac{\partial u_\varepsilon}{\partial t} - \frac{\partial}{\partial x}\left(K\left(\frac{x}{\varepsilon}\right)\frac{\partial u_\varepsilon}{\partial x}\right) = f(x,t), \quad x \in (0,1), t > 0,$$

$$(3.3.1)$$

with the boundary conditions

$$u_\varepsilon(0,t) = 0, \quad u_\varepsilon(1,t) = 0, \quad (3.3.2)$$

and the initial conditions

$$u_\varepsilon(x,0) = 0, \quad x \in (0,1). \quad (3.3.3)$$

To simplify the presentation of the homogenization method, assume that c, K, f, and ψ are smooth, scalar functions ($f \in C^2([0,1] \times [0,T])$) such that $c(\xi+1) = c(\xi)$ and $K(\xi+1) = K(\xi)$ for all real ξ, and c and K are positive. We assume that there exists $t_0 > 0$ such that $\forall t \in [0,t_0]$, $f(x,t) = 0$ and that $1/\varepsilon \in \mathbb{Z}$.

An asymptotic approximation is sought in the following form:

$$u_\varepsilon^{(2)} = u_0\left(x,\frac{x}{\varepsilon},t\right) + \varepsilon u_1\left(x,\frac{x}{\varepsilon},t\right) + \varepsilon^2 u_2\left(x,\frac{x}{\varepsilon},t\right), \quad (3.3.4)$$

where $u_j(x,\xi,t)$, $j = 0,1,2$, are smooth functions 1-periodic in ξ.

Substituting (3.3.4) in (3.3.1) and applying the chain-rule differentiation formula:

$$\frac{d}{dx}F\left(x,\frac{x}{\varepsilon}\right) = \left(\frac{\partial F}{\partial x}(x,\xi) + \varepsilon^{-1}\frac{\partial F}{\partial \xi}(x,\xi)\right)\big|_{\xi=\frac{x}{\varepsilon}},$$

then introducing the notation

$$L_{\alpha\beta} = \frac{\partial}{\partial \alpha}\left(K(\xi)\frac{\partial}{\partial \beta}\right), \quad \alpha,\beta \in \{x,\xi\},$$

and collecting the terms corresponding to powers of ε from -2 to 2, we get

$$L_\varepsilon u_\varepsilon^{(2)} = c\left(\frac{x}{\varepsilon}\right)\frac{\partial u_\varepsilon^{(2)}}{\partial t} - \frac{\partial}{\partial x}\left(K\left(\frac{x}{\varepsilon}\right)\frac{\partial u_\varepsilon^{(2)}}{\partial x}\right)$$

$$= \left\{ -\varepsilon^{-2}L_{\xi\xi}u_0(x,\xi,t) - \varepsilon^{-1}(L_{\xi\xi}u_1(x,\xi,t) + L_{\xi x}u_0(x,\xi,t)\right.$$

$$+ L_{x\xi}u_0(x,\xi,t)) - \varepsilon^0\left(L_{\xi\xi}u_2(x,\xi,t) + L_{\xi x}u_1(x,\xi,t)\right.$$

$$\left. + L_{x\xi}u_1(x,\xi,t) + L_{xx}u_0(x,\xi,t) - c(\xi)\frac{\partial u_0}{\partial t}(x,\xi,t)\right)$$

$$- \varepsilon^1\left(L_{\xi x}u_2(x,\xi,t) + L_{x\xi}u_2(x,\xi,t) + L_{xx}u_1(x,\xi,t)\right.$$

$$\left. - c(\xi)\frac{\partial u_1}{\partial t}(x,\xi,t)\right) - \varepsilon^2(L_{xx}u_2(x,\xi,t)$$

$$\left.\left. - c(\xi)\frac{\partial u_2}{\partial t}(x,\xi,t)\right)\right\}\bigg|_{\xi=\frac{x}{\varepsilon}}. \tag{3.3.5}$$

Here, the last two lines contain small terms (of order ε or ε^2). Variables x, ξ, t inside the brackets $\{,\}$ are considered as completely independent. The notation $L_{\alpha\beta}$ is introduced to make the expression of $L_\varepsilon u_\varepsilon^{(2)}$ shorter; for various values of α and β, it reads $L_{\xi\xi} = \frac{\partial}{\partial\xi}(K(\xi)\frac{\partial}{\partial\xi})$, $L_{\xi x} = \frac{\partial}{\partial\xi}(K(\xi)\frac{\partial}{\partial x})$, $L_{x\xi} = \frac{\partial}{\partial x}(K(\xi)\frac{\partial}{\partial\xi})$, $L_{xx} = \frac{\partial}{\partial x}(K(\xi)\frac{\partial}{\partial x})$.

Taking into account equation (3.3.1), we must asymptotically equate expansion (3.3.5) to the right-hand side f. This means that we require the terms of order ε^{-2} and ε^{-1} to vanish while the term of order ε^0 to be equal to $f(x,t)$. The terms of order ε or ε^2 constitute the discrepancy. Thus, we get three equations for the three 1-periodic in ξ terms $u_j(x,\xi,t)$, $j = 0,1,2$, of ansatz (3.3.4):

$$L_{\xi\xi}u_0(x,\xi,t) = 0, \tag{3.3.6_0}$$

$$L_{\xi\xi}u_1(x,\xi,t) + L_{\xi x}u_0(x,\xi,t) + L_{x\xi}u_0(x,\xi,t) = 0, \tag{3.3.6_1}$$

$$L_{\xi\xi}u_2(x,\xi,t) + L_{\xi x}u_1(x,\xi,t) + L_{x\xi}u_1(x,\xi,t) + L_{xx}u_0(x,\xi,t)$$

$$- c(\xi)\frac{\partial u_0}{\partial t}(x,\xi,t) = -f(x,t). \tag{3.3.6_2}$$

Solving equation (3.3.6$_0$), we get

$$u_0 = u_0(x, t), \qquad (3.3.7)$$

i.e. u_0 does not depend on ξ. Indeed, equation (3.3.6$_0$) reads

$$\frac{\partial}{\partial \xi}\left(K(\xi)\frac{\partial u_0}{\partial \xi}\right) = 0,$$

which means that $K(\xi)\frac{\partial u_0}{\partial \xi} = C_0(x, t)$ which is a constant with respect to ξ. So,

$$\frac{\partial u_0}{\partial \xi} = C_0(x, t)/K(\xi), \qquad (3.3.8)$$

where $\frac{\partial u_0}{\partial \xi}$ is a derivative of a periodic function. Let us apply now the following proposition.

Lemma 3.1. *Let F be a differentiable 1-periodic function of ξ. Then, $\langle \frac{\partial F}{\partial \xi}\rangle = 0$, where $\langle \cdot \rangle$ denotes the average over a period, i.e. integral $\int_0^1 \cdot \, d\xi$.*

Proof.

$$\left\langle \frac{\partial F}{\partial \xi}\right\rangle = \int_0^1 \frac{\partial F}{\partial \xi} d\xi = F(1) - F(0) = 0,$$

because F is a 1-periodic function. Lemma 3.1 is proved.
  Applying this lemma to (3.3.8), we get

$$0 = \left\langle \frac{\partial u_0}{\partial \xi}\right\rangle = \langle C_0(x, t)/K(\xi)\rangle = C_0(x, t)\left\langle \frac{1}{K(\xi)}\right\rangle,$$

and so $C_0(x, t) = 0$. Thus, $\frac{\partial u_0}{\partial \xi} = 0$, and so $u_0 = u_0(x, t)$.
  Now, consider equation (3.3.6$_1$), it reads

$$\frac{\partial}{\partial \xi}\left(K(\xi)\left(\frac{\partial u_1}{\partial \xi} + \frac{\partial u_0}{\partial x}\right)\right) = 0,$$

because $L_{x\xi}u_0(x,t) = 0$, and so

$$K(\xi)\left(\frac{\partial u_1}{\partial \xi} + \frac{\partial u_0}{\partial x}\right) = C_1(x,t),$$

which is a constant with respect to ξ. Repeating the above argument, we get

$$\frac{\partial u_1}{\partial \xi} = -\frac{\partial u_0}{\partial x} + C_1(x,t)\frac{1}{K(\xi)}, \qquad (3.3.9)$$

and so

$$0 = \left\langle \frac{\partial u_1}{\partial \xi} \right\rangle = \left\langle -\frac{\partial u_0}{\partial x} + C_1(x,t)\frac{1}{K(\xi)} \right\rangle = -\frac{\partial u_0}{\partial x} + C_1(x,t)\left\langle \frac{1}{K(\xi)} \right\rangle.$$

So,

$$C_1(x,t) = \left\langle \frac{1}{K(\xi)} \right\rangle^{-1} \frac{\partial u_0}{\partial x},$$

$$K(\xi)\left(\frac{\partial u_1}{\partial \xi} + \frac{\partial u_0}{\partial x}\right) = C_1(x,t) = \left\langle \frac{1}{K(\xi)} \right\rangle^{-1} \frac{\partial u_0}{\partial x}.$$

So,

$$K(\xi)\left(\frac{\partial u_1}{\partial \xi} + \frac{\partial u_0}{\partial x}\right) = \hat{K}\frac{\partial u_0}{\partial x}, \qquad (3.3.10)$$

where

$$\hat{K} = \left\langle \frac{1}{K(\xi)} \right\rangle^{-1}. \qquad (3.3.11)$$

Expressing now u_1 from (3.3.9), we determine it via function u_0.

Function u_1 is defined up to an arbitrary smooth function of x and t. Let us require that $u_1(x,0,t) = 0$. Then, due to the periodicity u_1 in ξ, we get $u_1(x,1/\varepsilon,t) = 0$ because $1/\varepsilon \in \mathbb{Z}$, and so $u_1(0,0,t) = u_1(1,1/\varepsilon,t) = 0$.

Finally, we consider equation $(3.3.6_2)$. If this equation has a periodic solution, then the average of the left-hand side is equal to the average of the right-hand side. Taking into account Lemma 3.1, we check that $\langle L_{\xi\xi}u_2(x,\xi,t)\rangle = 0$, $\langle L_{\xi x}u_1(x,\xi,t)\rangle = 0$, and so

$$\langle L_{x\xi}u_1(x,\xi,t) + L_{xx}u_0(x,t)\rangle - \langle c(\xi)\rangle\frac{\partial u_0}{\partial t}(x,t) = -f(x,t). \quad (3.3.12)$$

Solving completely (3.3.6$_2$), we can require for u_1 that $u_2(0,0,t) = u_2(1,1/\varepsilon,t) = 0$.

Applying now (3.3.10), (3.3.12) and taking into account that $L_{x\xi}u_1 + L_{xx}u_0 = \frac{\partial}{\partial x}(K(\xi)(\frac{\partial u_1}{\partial \xi} + \frac{\partial u_0}{\partial x}))$, we get the homogenized equation with respect to the unknown function $u_0(x,t)$:

$$\langle c(\xi)\rangle\frac{\partial u_0}{\partial t} - \hat{K}\frac{\partial^2 u_0}{\partial x^2} = f(x,t). \quad (3.3.13)$$

If u_0 satisfies this equation, then applying the same arguments to equations (3.3.6$_0$) and (3.3.6$_1$), we find the 1-periodic solution $u_2(x,\xi,t)$ of problem (3.3.6$_2$). We emphasize that (3.3.6$_2$), as well as (3.3.6$_0$), (3.3.6$_1$), are differential equations with respect to the fast variable ξ, while x and t are the real parameters.

Equation (3.3.13) describes the heat transfer in a homogeneous medium with heat capacity $\langle c(\xi)\rangle$ and conductivity $\hat{K} = \langle\frac{1}{K(\xi)}\rangle^{-1}$. The asymptotic approximation (3.3.4) satisfies the original equation (3.3.1) with the error of order ε and u_0 (solution to (3.3.13)) is the leading term of this asymptotic approximation. Imposing for u_0 the same boundary and initial conditions for u_ε,

$$u_0(0,t) = 0, \quad u_0(1,t) = 0, \quad (3.3.14)$$

$$u_0(x,0) = 0, \quad x \in (0,1), \quad (3.3.15)$$

we see that asymptotic approximation (3.3.4) satisfies exactly the boundary and initial conditions.

Problem (3.3.13)–(3.3.15) is called homogenized problem.

In the following section, we prove that

$$\|u_\varepsilon - u_0\|_{L^2((0,1)\times(0,T))} = O(\varepsilon), \quad (3.3.16)$$

$$\|u_\varepsilon - u_0 - \varepsilon u_1\|_{H^1((0,1)\times(0,T))} = O(\varepsilon). \quad (3.3.17)$$

Thus, the equivalent homogeneity hypothesis is justified in this example and the formulas for the macroscopic effective coefficients are

$\langle c(\xi) \rangle$ for the effective heat capacity and $\hat{K} = \langle \frac{1}{K(\xi)} \rangle^{-1}$ for the effective conductivity.

Note that in the multi-dimensional case, the algorithm for the computation of the effective coefficients is more complicated (see the following). For piecewise smooth coefficients, this analysis should be completed by a special treatment of the interface conditions (3.2.3), (3.2.4) adding the corresponding interface conditions for equation (3.3.6). The algorithm for the computation of effective coefficients is left without changes.

The heat equation (3.3.1) also describes some other physical processes, such as diffusion and flows in porous media.

4. Error Estimate

In the previous section, we constructed an asymptotic solution $u_\varepsilon^{(2)}$ such that

$$L_\varepsilon u_\varepsilon^{(2)} = f(x) + r_\varepsilon(x), \quad x \in (0,1), \tag{3.4.1}$$

where

$$r_\varepsilon(x) = \left\{ -\varepsilon \left(L_{\xi x} u_2(x, \xi, t) + L_{x\xi} u_2(x, \xi, t) \right. \right.$$

$$\left. + L_{xx} u_1(x, \xi, t) - c(\xi) \frac{\partial u_1}{\partial t}(x, \xi, t) \right) \tag{3.4.2}$$

$$\left. - \varepsilon^2 \left(L_{xx} u_2(x, \xi, t) - c(\xi) \frac{\partial u_2}{\partial t}(x, \xi, t) \right) \right\} \Big|_{\xi = \frac{x}{\varepsilon}},$$

see (3.3.5) and (3.3.6). Analyzing the construction of $u_\varepsilon^{(2)}$, we check that all terms in (3.4.2) are bounded, and so

$$\|r_\varepsilon\|_{L^2((0,1) \times (0,T)} = O(\varepsilon). \tag{3.4.3}$$

Subtract equation (3.3.1) from (3.4.1). Then, for the difference $w_\varepsilon = u_\varepsilon^{(2)} - u_\varepsilon$, we get

$$L_\varepsilon w_\varepsilon = r_\varepsilon(x), \quad x \in (0,1), \tag{3.4.4}$$

$$w_\varepsilon(0,t) = 0, \quad w_\varepsilon(1,t) = 0, \quad w_\varepsilon(x,0) = 0,$$

because $u^{(2)}$ satisfies the boundary and initial conditions exactly.

Then, applying *a priori* estimate (2.4.5) from Chapter 2, we get

$$\|w_\varepsilon\|_{H^1((0,1)\times(0,T))} \le C_D^p \|r_\varepsilon\|_{L^2((0,1)\times(0,T))}, \qquad (3.4.5)$$

where

$$C_D^p = \frac{\sqrt{1 + C_{PF}^2 + C_{PF}^4}}{\kappa}.$$

So,

$$\|w_\varepsilon\|_{H^1((0,1)\times(0,T))} = O(\varepsilon).$$

Note that

$$\frac{\partial u_2}{\partial x}(x, x/\varepsilon, t) = \varepsilon^{-1}\frac{\partial u_2}{\partial \xi} + \frac{\partial u_2}{\partial x},$$

so

$$\|u_2\|_{H^1((0,1)\times(0,T))} = O(\varepsilon^{-1}),$$

$$\|\varepsilon^2 u_2\|_{H^1((0,1)\times(0,T))} = O(\varepsilon).$$

So,

$$\|u_\varepsilon - u_0 - \varepsilon u_1\|_{H^1((0,1)\times(0,T))} = O(\varepsilon).$$

As

$$\|\varepsilon u_1\|_{L^2((0,1)\times(0,T))} = O(\varepsilon),$$

$$\|u_\varepsilon - u_0\|_{L^2((0,1)\times(0,T))} = O(\varepsilon).$$

Estimates (3.3.16), (3.3.17) are proved.

These define the order of the error in the homogenization method and describe the limits of the homogenization theory.

The described scheme is applied generally for the justification of the asymptotic methods.

Let us solve the problem

$$L_\varepsilon u_\varepsilon = f, \qquad (3.4.6)$$

where L_ε is some linear operator (including the interface, boundary, and initial conditions) such that for u_ε, we have the existence, uniqueness, and an *a priori* estimate

$$\|u_\varepsilon\| \le C_0 \varepsilon^{-\alpha} \|f\|, \tag{3.4.7}$$

where C_0 does not depend on ε.

We construct an approximate asymptotic solution u_ε^a satisfying (3.4.6) with a residual r_ε such that

$$\|r_\varepsilon\| = O(\varepsilon^K), \tag{3.4.8}$$

$$L_\varepsilon u_\varepsilon^a = f + r_\varepsilon. \tag{3.4.9}$$

Subtracting (3.4.6) from (3.4.9), we get for the difference

$$w = u_\varepsilon^a - u_\varepsilon,$$

$$L_\varepsilon w = r_\varepsilon. \tag{3.4.10}$$

Applying the *a priori* estimate (3.4.7), we get

$$\|w\| \le C_0 \varepsilon^{-\alpha} \|r_\varepsilon\| = O(\varepsilon^{K-\alpha}),$$

and we justify the *error estimate*

$$\|u_\varepsilon - u_\varepsilon^a\| = O(\varepsilon^{K-\alpha}). \tag{3.4.11}$$

A similar scheme is applicable in the case of nonlinear problems; however, the estimate (3.4.7) should be replaced with the stability estimate, evaluating the difference of the two solutions of problem (3.4.6) with two different right-hand sides.

Note that the approximate relations (3.4.8), (3.4.9) ensure the justification at the physical level, while mathematical results normally require estimates of (3.4.11) type or at least the proof of convergence if the method does not allow to get the error estimates.

Remark 4.1. We have presented the version of the homogenization method based on the asymptotic expansions. The alternative versions are the so-called G-convergence, Γ-convergence, two-scale convergence, and unfolding. These versions give as the result the convergence of u_ε to the solution u_0 of the homogenized problem as $\varepsilon \to 0$. The convergence methods do not require the regularity of the right-hand side, but usually they give less information about the solution than the asymptotic expansion method. One can find the description of these methods in Refs. [6, 7, 9, 12, 13, 18, 19, and 21].

5. Homogenization: Multiple Dimensions

In the case of multiple dimensions, the homogenization method works similarly to the 1-D case.

Consider the stationary equation

$$-\sum_{i,j=1}^{d} \frac{\partial}{\partial x_i} \left(A_{ij} \left(\frac{x}{\varepsilon} \right) \frac{\partial \mathbf{u}_\varepsilon}{\partial x_j} \right) = \mathbf{f}(x), \quad x \in G, \qquad (3.5.1)$$

with the Dirichlet's boundary condition

$$\mathbf{u}_\varepsilon \big|_{\partial G} = 0. \qquad (3.5.2)$$

We consider the following two versions of this problem: conductivity equation (C) and elasticity equation (E).

In case (C), the unknown function u_ε is a scalar function, as is the right-hand side f. In case (E), $\mathbf{u}_\varepsilon$ and $\mathbf{f}$ are vector-valued functions. The coefficients $A_{ij}(\xi)$ in both cases are piecewise smooth 1-periodic functions of the fast variable ξ:

$$\forall \xi \in \mathbb{R}^d, \quad \forall \mathbf{z} \in \mathbb{Z}^d, \quad A_{ij}(\xi + \mathbf{z}) = A_{ij}(\xi).$$

In case (C), every A_{ij} is a scalar function, while in case (E), it is a $d \times d$ matrix-valued function.

In both cases, A_{ij} satisfy conditions (i) and (ii) in Sections 4 and 5 of Chapter 2:

— in case (C):
(i) $\forall \xi \in \mathbb{R}^d, \quad A_{ij}(\xi) = A_{ji}(\xi)$,
(ii) $\exists \kappa > 0$ such that $\forall \xi \in \mathbb{R}^d, \forall \eta = (\eta_1, \ldots, \eta_d) \in \mathbb{R}^d$,

$$\sum_{i,j=1}^{d} A_{ij}(\xi) \eta_j \eta_i \geq \kappa \sum_{j=1}^{d} \eta_j^2.$$

— in case (E):
(i) $\forall \xi \in \mathbb{R}^d, \quad a_{ij}^{kl}(\xi) = a_{ji}^{lk}(\xi) = a_{kj}^{il}(\xi)$,
(ii) $\exists \kappa > 0$ such that $\forall \xi \in \mathbb{R}^d, \forall \eta \in \mathbb{R}^{d \times d}_{\text{sym}}$,

$$\sum_{i,j,k,l=1}^{d} a_{ij}^{kl}(\xi) \eta_{jl} \eta_{ik} \geq \kappa \sum_{j,l=1}^{d} \eta_{jl}^2.$$

In both cases, G is a bounded domain with the $C^{(4)}$-smooth boundary and $\mathbf{f} \in C^{(2)}(\bar{G})$.

The construction of the asymptotic approximation is similar in both cases, so we use the "vectorial" notation of problem (3.5.1), (3.5.2) corresponding to case (E).

As in the previous section, first for simplicity, we assume that A_{ij} are $C^{(1)}$ smooth functions, but later in Remark 5.1, we will describe the modifications which are needed in the case of piecewise $C^{(1)}$-smooth coefficients.

Similar to the previous section, an asymptotic approximation is sought in the form

$$\mathbf{u}_\varepsilon^{(2)} = \mathbf{u}_0\left(x, \frac{x}{\varepsilon}\right) + \varepsilon \mathbf{u}_1\left(x, \frac{x}{\varepsilon}\right) + \varepsilon^2 \mathbf{u}_2\left(x, \frac{x}{\varepsilon}\right), \qquad (3.5.3)$$

where $\mathbf{u}_j(x, \xi)$, $j = 0, 1, 2$ are smooth 3-D vector-valued functions 1-periodic in ξ in case (E), while in case (C), they are scalars.

Plugging (3.5.3) in (3.5.1), applying the chain-rule differentiation formula,

$$\frac{d}{dx_i} F\left(x, \frac{x}{\varepsilon}\right) = \left(\frac{\partial F}{\partial x_i}(x, \xi) + \varepsilon^{-1} \frac{\partial F}{\partial \xi_i}(x, \xi)\right)\Big|_{\xi = \frac{x}{\varepsilon}},$$

introducing notation

$$L_{\alpha\beta} = \sum_{i,j=1}^{d} \frac{\partial}{\partial \alpha_i}\left(A_{ij}(\xi) \frac{\partial}{\partial \beta_j}\right), \qquad \alpha, \beta \in \{x, \xi\},$$

and $L_\varepsilon = -\sum_{i,j=1}^{d} \frac{\partial}{\partial x_i}(A_{ij}(\frac{x}{\varepsilon}) \frac{\partial}{\partial x_j})$, collecting the terms corresponding to powers of ε from -2 to 2, we get as in the previous section

$$L_\varepsilon \mathbf{u}_\varepsilon^{(2)} = \Big\{ -\varepsilon^{-2} L_{\xi\xi} \mathbf{u}_0(x, \xi) - \varepsilon^{-1}(L_{\xi\xi} \mathbf{u}_1(x, \xi) + L_{\xi x} \mathbf{u}_0(x, \xi)$$

$$+ L_{x\xi} \mathbf{u}_0(x, \xi)) - \varepsilon^0(L_{\xi\xi} \mathbf{u}_2(x, \xi)$$

$$+ L_{\xi x} \mathbf{u}_1(x, \xi) + L_{x\xi} \mathbf{u}_1(x, \xi) + L_{xx} \mathbf{u}_0(x, \xi))$$

$$- \varepsilon^1(L_{\xi x} \mathbf{u}_2(x, \xi) + L_{x\xi} \mathbf{u}_2(x, \xi) + L_{xx} \mathbf{u}_1(x, \xi))$$

$$- \varepsilon^2(L_{xx} \mathbf{u}_2(x, \xi)) \Big\}\Big|_{\xi = \frac{x}{\varepsilon}}. \qquad (3.5.4)$$

Here, the last two lines contain small terms (of order ε or ε^2). Denote it by

$$r_\varepsilon(x) = \Big\{ - \varepsilon^1 (L_{\xi x} \mathbf{u}_2(x, \xi) + L_{x\xi} \mathbf{u}_2(x, \xi) + L_{xx} \mathbf{u}_1(x, \xi))$$

$$- \varepsilon^2 L_{xx} \mathbf{u}_2(x, \xi) \Big\}|_{\xi = \frac{x}{\varepsilon}}. \tag{3.5.4$_r$}$$

The variables x, ξ inside the brackets $\{\ \}$ are considered as completely independent.

Taking into account equation (3.5.1), we must asymptotically equate expansion (3.5.4) to the right-hand side $\mathbf{f}$. It means that we require the terms of order ε^{-2} and ε^{-1} to be equal to zero while the term of order ε^0 to be equal to $\mathbf{f}(x)$. The terms of order ε or ε^2 constitute the discrepancy. Thus, we get three equations for the three 1-periodic in ξ terms $\mathbf{u}_j(x, \xi)$, $j = 0, 1, 2$, of ansatz (3.5.3):

$$L_{\xi\xi} \mathbf{u}_0(x, \xi) = \mathbf{0}, \tag{3.5.5$_0$}$$

$$L_{\xi\xi} \mathbf{u}_1(x, \xi) + L_{\xi x} \mathbf{u}_0(x, \xi) + L_{x\xi} \mathbf{u}_0(x, \xi) = \mathbf{0}, \tag{3.5.5$_1$}$$

$$L_{\xi\xi} \mathbf{u}_2(x, \xi) + L_{\xi x} \mathbf{u}_1(x, \xi) + L_{x\xi} \mathbf{u}_1(x, \xi) + L_{xx} \mathbf{u}_0(x, \xi) = - \mathbf{f}(x). \tag{3.5.5$_2$}$$

We use the following theorem.

Theorem 5.1. *Let* $\mathbf{F}$ *be a piecewise continuous 1-periodic vector-valued function. Then, the equation*

$$L_{\xi\xi} \mathbf{U}(\xi) = \mathbf{F}(\xi)$$

admits a unique (up to an additive constant vector) 1-periodic solution iff $\langle \mathbf{F} \rangle = \mathbf{0}$. *Any solution* $\mathbf{U}$ *is a sum* $\mathbf{U}_0 + \mathbf{C}$, *where* $\mathbf{U}_0$ *is a unique solution with the vanishing mean value* $\langle \mathbf{U}_0 \rangle = \mathbf{0}$ *and* $\mathbf{C}$ *is an arbitrary constant.*

Here, $\langle \cdot \rangle$ denotes the average over a period, i.e. the integral $\int_0^1 \cdots \int_0^1 \cdot \, d\,\xi_1 \cdots d\,\xi_d$.

Note that the piecewise continuity of the right-hand side in this theorem may be replaced by belonging to the class $L_\sharp^2$, where $L_\sharp^2$ is the space of 1-periodic functions belonging to $L^2(B)$ for any ball $B \subset \mathbb{R}^d$.

This theorem can be proved in the same way as Theorem 4.1 in Chapter 2 using the Riesz–Fréchet representation theorem. The only one modification is related to the periodicity conditions for the equation

$$-\sum_{i,j=1}^{d} \frac{\partial}{\partial \xi_i} \left(A_{ij}(\xi) \frac{\partial \mathbf{U}}{\partial \xi_j} \right) = -\mathbf{F}(\xi).$$

Namely, we introduce the Sobolev space $H_\sharp^1$ as the completion of the space $C_\sharp^{(\infty)}$ of 1-periodic functions of $C^{(\infty)}(\mathbb{R}^d)$ with respect to the inner product of $H^1((0,1)^d)$. We also introduce the subspace $H_{\sharp,0}^1$ of functions of $H_\sharp^1$ having the vanishing mean value $\langle \, \cdot \, \rangle$. We define a weak solution $\mathbf{U}$ as a vector-valued function of $H_\sharp^1$ satisfying $\forall \mathbf{V} \in H_\sharp^1$ the integral identity

$$\left\langle \sum_{i,j=1}^{d} A_{ij}(\xi) \frac{\partial \mathbf{U}}{\partial \xi_j} \cdot \frac{\partial \mathbf{V}}{\partial \xi_i} \right\rangle = \langle -\mathbf{F} \cdot \mathbf{V} \rangle.$$

Using the Poincaré inequality and Korn inequality, we check that the left-hand side of the integral defines an inner product on $H_{\sharp,0}^1$ and, applying the Riesz–Fréchet theorem, prove that there exists a unique $\mathbf{U}_0 \in H_{\sharp,0}^1$ such that $\forall \mathbf{V} \in H_{\sharp,0}^1$,

$$\left\langle \sum_{i,j=1}^{d} A_{ij}(\xi) \frac{\partial \mathbf{U}_0}{\partial \xi_j} \cdot \frac{\partial \mathbf{V}}{\partial \xi_i} \right\rangle = \langle -\mathbf{F} \cdot \mathbf{V} \rangle.$$

Finally, using condition $\langle \mathbf{F} \rangle = 0$, we check that this integral identity remains valid for all $\mathbf{V} \in H_\sharp^1$ (note that any function $\mathbf{V}$ of $H_\sharp^1$ can be presented in the form

$$\mathbf{V} = \mathbf{V}_0 + \langle \mathbf{V} \rangle, \quad \mathbf{V}_0 \in H_{\sharp,0}^1).$$

Solving equation (3.5.5$_0$), it is evident that the zero function is its solution, so according to Theorem 5.1, all its solutions are constant (with respect to the ξ variable). So, the set of solutions is the set of functions depending on x variable only:

$$\mathbf{u}_0 = \mathbf{u}_0(x), \tag{3.5.6}$$

i.e. $\mathbf{u}_0$ does not depend on ξ.

So, in equation $(3.5.5_1)$, the last term vanishes. Similar to Section 3, we have the following lemma.

Lemma 5.1. *Let F be a differentiable 1-periodic function of ξ. Then,*
$\langle \frac{\partial F}{\partial \xi_i} \rangle = 0$.

Proof. According to Fubbini theorem, the order of integration in the integral $\langle \frac{\partial F}{\partial \xi_i} \rangle$ with respect to $\xi_1, \xi_2, \ldots, \xi_d$ may be changed so that the first integration be held in ξ_i. Then, applying the same arguments as in the proof of Lemma 2.1, we get that this integration with respect to ξ_i gives zero. So, the next $d - 1$ integrations are applied to the function zero and this proves the lemma. $\qquad\square$

Applying this lemma to the second term of $(3.5.5_1)$, we see that the necessary and sufficient condition for the existence of a 1-periodic solution is satisfied. Moreover, this second term has the form

$$\sum_{l=1}^{d} \left(\sum_{i=1}^{d} \frac{\partial}{\partial \xi_i} A_{il}(\xi) \right) \frac{\partial u_0}{\partial x_l}(x), \qquad (3.5.7)$$

i.e. it is a linear combination of d functions depending on ξ: $(\sum_{i=1}^{d} \frac{\partial}{\partial \xi_i} A_{il}(\xi))$. This gives an idea to seek a solution for $(3.5.5_1)$ in the form of the linear combination

$$\mathbf{u}_1(x, \xi) = \sum_{l=1}^{d} N_l(\xi) \frac{\partial u_0}{\partial x_l}(x), \qquad (3.5.8)$$

according to the superposition principle [22]. Here, $N_l, l = 1, 2, ..., d$ are d unknown $d \times d$ matrix-valued functions, 1-periodic in ξ (in case (C), N_l are scalar functions). They satisfy the so-called *cell problems*: Find N_l satisfying equation

$$L_{\xi\xi} N_l = -\sum_{i=1}^{d} \frac{\partial}{\partial \xi_i} A_{il}(\xi). \qquad (3.5.9)$$

One can check directly that the function (3.5.8) with N_l satisfying (3.5.9) is a solution to the problem $(3.5.5_1)$ with the right-hand side (3.5.7).

Consider finally, the equation $(3.5.5_2)$:

$$L_{\xi\xi}\mathbf{u}_2(x,\xi) = -L_{\xi x}\mathbf{u}_1(x,\xi) - L_{x\xi}\mathbf{u}_1(x,\xi) - L_{xx}\mathbf{u}_0(x) - \mathbf{f}(x). \quad (3.5.5'_2)$$

According to Theorem 5.1, the necessary and sufficient condition for the existence of 1-periodic solution to this equation is

$$\langle -L_{\xi x}\mathbf{u}_1(x,\xi) - L_{x\xi}\mathbf{u}_1(x,\xi) - L_{xx}\mathbf{u}_0(x,\xi) - \mathbf{f}(x) \rangle = 0, \quad (3.5.10)$$

where $\langle -L_{\xi x}\mathbf{u}_1(x,\xi) \rangle = 0$ due to Lemma 5.1, $\langle \mathbf{f}(x) \rangle = \mathbf{f}(x)$ and

$$\langle L_{x\xi}\mathbf{u}_1(x,\xi) + L_{xx}\mathbf{u}_0(x) \rangle$$

$$= \left\langle \sum_{i,j=1}^{d} \frac{\partial}{\partial x_i} \left(A_{ij}(\xi) \frac{\partial(\sum_{l=1}^{d} N_l(\xi) \frac{\partial \mathbf{u}_0}{\partial x_l})}{\partial \xi_j} \right) \right.$$

$$\left. + \sum_{i,l=1}^{d} \frac{\partial}{\partial x_i} \left(A_{il}(\xi) \frac{\partial \mathbf{u}_0}{\partial x_l} \right) \right\rangle$$

$$= \sum_{i,l=1}^{d} \frac{\partial}{\partial x_i} \left(\left\langle \sum_{j=1}^{d} A_{ij}(\xi) \frac{\partial N_l(\xi)}{\partial \xi_j} + A_{il}(\xi) \right\rangle \frac{\partial \mathbf{u}_0}{\partial x_l} \right)$$

$$= \sum_{i,l=1}^{d} \frac{\partial}{\partial x_i} \left(\hat{A}_{il} \frac{\partial \mathbf{u}_0}{\partial x_l} \right),$$

where $\hat{A}_{il}$ are defined by the formula

$$\hat{A}_{il} = \left\langle \sum_{j=1}^{d} A_{ij}(\xi) \frac{\partial N_l(\xi)}{\partial \xi_j} + A_{il} \right\rangle. \quad (3.5.11)$$

Thus, the necessary and sufficient condition for the existence of a 1-periodic solution to equation $(3.5.5_2)$ is the equation

$$-\sum_{i,l=1}^{d} \frac{\partial}{\partial x_i} \left(\hat{A}_{il} \frac{\partial \mathbf{u}_0}{\partial x_l} \right) = \mathbf{f}(x), \quad x \in G. \quad (3.5.12)$$

It can be proved that $\hat{A}_{il}$ satisfy relations **(i)** and **(ii)** (see Ref. [3] and Remark 5.2). If $\mathbf{u}_0$ is a solution to (3.5.12), then we can solve numerically equation $(3.5.5_2)$ and find $\mathbf{u}_2$ such that equation $(3.5.1)$ is satisfied up to a residual of order ε. Moreover, if $\mathbf{u}_0$ satisfies the condition

$$\mathbf{u}_0|_{\partial\Omega} = 0, \tag{3.5.13}$$

then the condition $(3.5.2)$ is also satisfied up to a residual of order ε. The leading term $\mathbf{u}_0$ of the asymptotic approximation $\mathbf{u}_\varepsilon^{(2)}$ is a solution for the problem $(3.5.12)$, $(3.5.13)$ of the same type as the problem $(3.5.1)$, $(3.5.2)$ but describes a homogeneous medium with the macroscopic (effective) coefficients $\hat{A}_{il}$. It justifies the hypothesis of the equivalent homogeneity for the elasticity equation and the algorithm $(3.5.9)$, $(3.5.11)$, $(3.5.12)$, $(3.5.13)$ for the computation of the macroscopic constants.

However, the most important for applications are the stresses and strains (the failure criteria depend on these values). In order to obtain their approximations, one should use the first corrector approximation, i.e.

$$\mathbf{u}^{(1)}(x) = \left(\mathbf{u}_0 + \varepsilon \sum_{l=1}^{d} N_l(\xi) \frac{\partial \mathbf{u}_0}{\partial x_l}(x) \right) \Bigg|_{\xi = x/\varepsilon}.$$

The corresponding approach can be found in Chapter 5 of Ref. [3]: the stresses σ_{ik} can be approximately calculated as

$$\sigma_{ik} = \sum_{j,l=1}^{d} a_{ij}^{kl}\left(\frac{x}{\varepsilon}\right) e_{jl}(\mathbf{u}^{(1)}).$$

Remark 5.1. If the coefficients A_{ij} are discontinuous at some smooth surfaces being globally piecewise smooth functions, then at these surfaces, the equation should be replaced by two interface conditions: the continuity of the displacement vector

$$[\mathbf{u}_\varepsilon] = 0, \tag{3.5.14}$$

and the continuity of the normal stress vector

$$\left[\sum_{i,j=1}^{d} n_i \left(A_{ij}\left(\frac{x}{\varepsilon}\right) \frac{\partial \mathbf{u}_\varepsilon}{\partial x_j} \right) \right] = 0, \tag{3.5.15}$$

where the square brackets [ ] denote the jump of the function in transition through the interface and $\mathbf{n} = (n_1, n_2, n_3)$ (for the case $d = 3$) stands for the normal vector to the interface. (Another approach is concerned with the weak formulation of the problem (3.5.1), (3.5.2) obtained by the multiplication of the equation (3.5.1) by a test function and then by integration over G by parts. Then, the conditions (3.5.14) and (3.5.15) can be removed.) In this case, the analysis should be completed by a special treatment of the interface conditions (3.5.14), (3.5.15) adding the corresponding interface conditions for equation (3.5.5). Namely,

$$[\mathbf{u}_l] = \mathbf{0}, l = 0, 1, 2, \tag{3.5.16}$$

and

$$\left[\sum_{i,j=1}^{d} n_i(\xi)(A_{ij}(\xi)) \left(\frac{\partial \mathbf{u}_l}{\partial \xi_j} + \frac{\partial \mathbf{u}_{l-1}}{\partial x_j} \right) \right] = \mathbf{0} \tag{3.5.17}$$

(for $l = 0$, the last term should be omitted). The algorithm for the computation of effective coefficients is left without changes.

Remark 5.2. Let us prove that the homogenized coefficients $\hat{A}_{il}$ satisfy the properties (**i**) and (**ii**).

Theorem 5.2. *In case* (C), $\hat{A}_{il}$ *satisfy:*

(**i**) $\hat{A}_{il} = \hat{A}_{li}, \quad i, l \in \{1, \ldots, d\}$,
(**ii**) $\forall \eta = (\eta_1, \ldots, \eta_d) \in \mathbb{R}^d$.

$$\sum_{i,l=1}^{d} \hat{A}_{il} \eta_l \eta_i \geq \kappa \sum_{l=1}^{d} \eta_l^2.$$

In case (E), $\hat{A}_{il}$ *satisfy:*

(**i**) $\hat{a}_{il}^{sp} = \hat{a}_{li}^{ps} = \hat{a}_{sl}^{ip}, \quad i, l, s, p \in \{1, \ldots, d\}$,
(**ii**) $\forall \eta \in \mathbb{R}_{\text{sym}}^{d \times d} \sum_{i,l,s,p=1}^{d} \hat{a}_{il}^{sp} \eta_{lp} \eta_{is} \geq \kappa \sum_{l,p=1}^{d} (\eta_{lp})^2.$

Proof. Let us start with case (C).

Rewrite the cell problems for N_l in the form

$$L_{\xi\xi} N_l + \sum_{i=1}^{d} \frac{\partial}{\partial \xi_i} A_{il}(\xi) = 0,$$

i.e.

$$L_{\xi\xi}(N_l + \xi_l) = 0. \tag{3.5.18}$$

Denote

$$M_l = N_l + \xi_l. \tag{3.5.19}$$

The weak solution is defined as M_l such that $M_l(\xi) - \xi_l \in H_{\sharp}^1$ and for all $\varphi \in H_{\sharp}^1$,

$$\left\langle \sum_{i,j=1}^{d} A_{ij} \frac{\partial M_l}{\partial \xi_j} \frac{\partial \varphi}{\partial \xi_i} \right\rangle = 0. \tag{3.5.20}$$

On the other hand,

$$\hat{A}_{il} = \left\langle \sum_{j=1}^{d} A_{ij} \frac{\partial M_l}{\partial \xi_j} \right\rangle, \tag{3.5.21}$$

$$\hat{A}_{il} = \left\langle \sum_{k,j=1}^{d} A_{kj} \frac{\partial M_l}{\partial \xi_j} \frac{\partial \xi_i}{\partial \xi_k} \right\rangle. \tag{3.5.22}$$

Adding to this expression, a particular case of the variational identity (3.5.20) ($\varphi = N_l$)

$$\left\langle \sum_{k,j=1}^{d} A_{kj} \frac{\partial M_l}{\partial \xi_j} \frac{\partial N_i}{\partial \xi_k} \right\rangle = 0,$$

we get

$$\hat{A}_{il} = \left\langle \sum_{k,j=1}^{d} A_{kj} \frac{\partial M_l}{\partial \xi_j} \frac{\partial M_i}{\partial \xi_k} \right\rangle. \tag{3.5.23}$$

Now (**i**) is evident. Let us prove (**ii**) for $\hat{A}_{il}$.

Consider

$$\sum_{i,l=1}^{d} \hat{A}_{il}\eta_l\eta_i = \left\langle \sum_{i,l,k,j=1}^{d} A_{kj} \frac{\partial(M_l\eta_l)}{\partial\xi_j} \frac{\partial(M_i\eta_i)}{\partial\xi_k} \right\rangle$$

$$\geq \left\langle \kappa \sum_{j=1}^{d} \left(\frac{\partial\left(\sum_{l=1}^{d} M_l\eta_l\right)}{\partial\xi_j} \right)^2 \right\rangle$$

$$\geq \kappa \sum_{j=1}^{d} \left\langle \frac{\partial}{\partial\xi_j} \left(\sum_{l=1}^{d} M_l\eta_l \right) \right\rangle^2$$

$$= \kappa \sum_{j=1}^{d} \left\langle \frac{\partial}{\partial\xi_j} \sum_{l=1}^{d} (N_l\eta_l + \xi_l\eta_l) \right\rangle^2$$

$$\geq \kappa \sum_{l=1}^{d} \eta_l^2 \qquad\qquad (3.5.24)$$

because, due to Lemma 5.1,

$$\left\langle \frac{\partial}{\partial\xi_j} \left(\sum_{l=1}^{d} N_l\eta_l \right) \right\rangle = 0.$$

Property (**ii**) is proved.

Consider case (E).
In this case, the cell problem takes the form

$$L_{\xi\xi}(N_l + \xi_l I) = 0. \qquad\qquad (3.5.25)$$

Here I is the identity matrix. Denote $M_l = N_l + \xi_l I$.
The weak solution is defined as a matrix-valued function M_l such that $M_l(\xi) - \xi_l I \in (H^1_\#)^{d\times d}$ and for all $\varphi \in (H^1_\#)^{d\times d}$,

$$\left\langle \sum_{i,j=1}^{d} \frac{\partial\varphi^T}{\partial\xi_i} A_{ij} \frac{\partial M_l}{\partial\xi_j} \right\rangle = 0. \qquad\qquad (3.5.26)$$

Here, $(H^1_\#)^{d\times d}$ is the space of $d \times d$ matrix-valued functions with entries from $H^1_\#$.

Then, as in case (C), we get

$$\hat{A}_{il} = \left\langle \sum_{k,j=1}^{d} \left(\frac{\partial M_i}{\partial \xi_k}\right)^T A_{kj} \frac{\partial M_l}{\partial \xi_j} \right\rangle. \qquad (3.5.27)$$

So, $\hat{A}_{il} = \hat{A}_{li}^T$. Denote $\hat{a}_{il}^{sp}$ as the entries of $\hat{A}_{il}$. We get

$$\hat{a}_{il}^{sp} = \hat{a}_{li}^{ps}. \qquad (3.5.28)$$

On the other hand,

$$\hat{A}_{il} = \left\langle \sum_{j=1}^{d} A_{ij} \frac{\partial M_l}{\partial \xi_j} \right\rangle,$$

so using (**i**) for A_{ij}, we get

$$\hat{a}_{il}^{sp} = \left\langle \sum_{j,m=1}^{d} a_{ij}^{sm} \frac{\partial M_l^{mp}}{\partial \xi_j} \right\rangle$$

$$= \left\langle \sum_{j,m=1}^{d} a_{sj}^{im} \frac{\partial M_l^{mp}}{\partial \xi_j} \right\rangle$$

$$= \hat{a}_{sl}^{ip}. \qquad (3.5.29)$$

Here, M_l^{mp} are the entries of matrices M_l. So, (**i**) is proved.

Let η be a symmetric matrix with entries η_{lp}. Consider the form

$$\mathcal{I} = \sum_{i,l,s,p=1}^{d} \hat{a}_{il}^{sp} \eta_{lp} \eta_{is}. \qquad (3.5.30)$$

Denote N_l^{mp} the entries of the matrix N_l. Then, $M_l^{mp} = N_l^{mp} + \xi_l \delta_{mp}$.

The formula (3.5.27) can be presented in the form

$$\hat{a}_{il}^{sp} = \sum_{j,m,q,r=1}^{d} \left\langle a_{qj}^{rm} e_{jm}(\mathbf{M}_l^p) e_{qr}(\mathbf{M}_i^s) \right\rangle, \qquad (3.5.31)$$

where $\mathbf{M}_l^p$ is the pth column of the matrix M_l and

$$e_{jm}(\mathbf{u}) = \frac{1}{2}\left(\frac{\partial u_m}{\partial \xi_j} + \frac{\partial u_j}{\partial \xi_m}\right).$$

Then,

$$\mathcal{I} = \sum_{i,l,s,p,j,m,q,r=1}^{d} \left\langle a_{qj}^{rm} e_{jm} \left(\mathbf{M}_l^p \eta_{lp} \right) e_{qr} \left(\mathbf{M}_i^s \eta_{is} \right) \right\rangle$$

$$= \sum_{j,m,q,r=1}^{d} \left\langle a_{qj}^{rm} e_{jm} \left(\sum_{l,p=1}^{d} \mathbf{M}_l^p \eta_{lp} \right) e_{qr} \left(\sum_{i,s=1}^{d} \mathbf{M}_i^s \eta_{is} \right) \right\rangle$$

$$\geq \kappa \sum_{j,m=1}^{d} \left\langle \left\{ e_{jm} \left(\sum_{l,p=1}^{d} \mathbf{M}_l^p \eta_{lp} \right) \right\}^2 \right\rangle$$

$$\geq \kappa \sum_{j,m=1}^{d} \left\langle e_{jm} \left(\sum_{l,p=1}^{d} \mathbf{M}_l^p \eta_{lp} \right) \right\rangle^2$$

$$= \kappa \sum_{j,m=1}^{d} \left\langle e_{jm} \left(\sum_{l,p=1}^{d} \xi_l \mathbf{e}_p \eta_{lp} \right) \right\rangle^2,$$

where $\mathbf{e}_p = (\delta_{pt})_{1 \leq t \leq d}^T$.

Then, the last expression is equal to

$$\kappa \sum_{j,m,l,p=1}^{d} \left\langle \frac{1}{2} \left(\delta_{jl} \delta_{mp} \eta_{lp} + \delta_{jp} \delta_{ml} \eta_{lp} \right) \right\rangle^2 = \kappa \sum_{l,p=1}^{d} \eta_{lp}^2.$$

So,

$$\mathcal{I} \geq \kappa \sum_{l,p=1}^{d} \eta_{lp}^2. \tag{3.5.32}$$

Thus, **(ii)** is proved.   □

The same analysis can be applied to the time-dependent elasticity equation

$$\rho \left(\frac{x}{\varepsilon} \right) \frac{\partial^2 \mathbf{u}_\varepsilon}{\partial t^2} - \sum_{i,j=1}^{d} \frac{\partial}{\partial x_i} \left(A_{ij} \left(\frac{x}{\varepsilon} \right) \frac{\partial \mathbf{u}_\varepsilon}{\partial x_j} \right) = \mathbf{f}(x,t),$$

where $\rho(\frac{x}{\varepsilon}) > 0$ is the density. The homogenized equation has the form

$$\hat{\rho}\frac{\partial^2 \mathbf{u}_0}{\partial t^2} - \sum_{i,j=1}^{d} \frac{\partial}{\partial x_i}\left(\hat{A}_{ij}\frac{\partial \mathbf{u}_0}{\partial x_j}\right) = \mathbf{f}(x,t),$$

where $\hat{\rho} = <\rho>$. This means that the density is homogenized by a simple averaging, while the elasticity moduli having a more complicated homogenization algorithm (3.5.9), (3.5.11).

As for the history of homogenization, let us mention the pioneering works on the asymptotic analysis of equations with rapidly oscillating coefficients by E. Sanchez-Palencia [20], E. De Giorgi and S. Spagnollo [7], N. Bakhvalov [2], J. L. Lions, A. Bensoussan, and G. Papanicolaou [4], V. Jikov, S. Kozlov, O. Oleinik [9], V. Berdichevsky [24], I. Babuska [1], V. Marchenko and E. Khruslov [11], F. Murat [13], and L. Tartar [21].

The above homogenization approach can be applied to various equations describing processes in periodic micro-heterogeneous media: the heat equation, elasticity equation, Navier–Stokes equations in porous media, and so on (see Refs. [3–6, 12, 14, 17, 18, 20, and 19]).

Concluding this short introduction to the up-scaling process for periodic structures, we discuss the applicability of other methods to this process. One of the most popular methods in physics for the passage from one scale to another among the quantum field theories is the renormalization group method (see Refs. [8, 23], and their bibliographies). The renormalization group method assumes existence of a great (infinite) number of scales in the physical process such that the process is self-similar with respect to these scales. This is the case for some stochastic models (of say, statistical physics) but not the case of the periodic heterogeneous media. The applicability of the renormalization group theory to the problems for the periodic micro-structures is still an open question.

6. Error Estimate: Multiple Dimensions

As in Section 3, we see that

$$L_\varepsilon \mathbf{u}_\varepsilon^{(2)} = \mathbf{f}(x) + \mathbf{r}_\varepsilon(x), \tag{3.6.1}$$

where

$$\|\mathbf{r}_\varepsilon\|_{L^2(G)} = O(\varepsilon). \tag{3.6.2}$$

However, the boundary condition for u_ε is satisfied with the residual of order ε:

$$\mathbf{u}_\varepsilon^{(2)}\big|_{\partial G} = \varepsilon\mathbf{u}_1(x, x/\varepsilon)|_{\partial G} + \varepsilon^2\mathbf{u}_2(x, x/\varepsilon).$$

Let us modify the asymptotic approximation $\mathbf{u}_\varepsilon^{(2)}$ and consider

$$\tilde{\mathbf{u}}_\varepsilon^{(2)} = \mathbf{u}_\varepsilon^{(2)} + (1 - \chi_\varepsilon(x))\left(\varepsilon\mathbf{u}_1(x, x/\varepsilon) + \varepsilon^2\mathbf{u}_2(x, x/\varepsilon)\right), \tag{3.6.3}$$

where

$$\chi_\varepsilon(x) = \begin{cases} 1 \text{ in } G\backslash G_\varepsilon, \\ \varepsilon^{-1} \text{ dist } (x, \partial G) \text{ in } G_\varepsilon, \end{cases}$$

G_ε was defined in Chapter 1 as the ε-neighborhood of ∂G within the domain G.

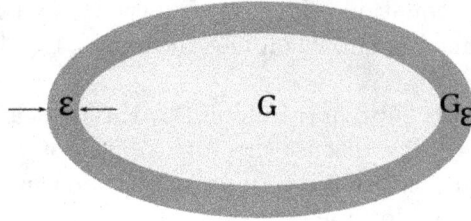

Note that

$$\mathrm{supp}(1 - \chi_\varepsilon) \subset G_\varepsilon, \tag{3.6.4}$$

$$|1 - \chi_\varepsilon(x)| \le 1, \quad |\nabla(1 - \chi_\varepsilon(x))| \le C\varepsilon^{-1}. \tag{3.6.5}$$

Then,

$$L_\varepsilon\tilde{\mathbf{u}}_\varepsilon^{(2)} = \mathbf{f}(x) + \mathbf{r}_\varepsilon(x) - \sum_{i=1}^{d} \frac{\partial}{\partial x_i}\mathbf{r}_{\varepsilon i}(x), \tag{3.6.6}$$

where

$$\mathbf{r}_{\varepsilon i}(x) = -\sum_{j=1}^{d} A_{ij}\left(\frac{x}{\varepsilon}\right) \frac{\partial}{\partial x_j}\left\{(1 - \chi_\varepsilon(x))(\varepsilon\mathbf{u}_1 + \varepsilon^2\mathbf{u}_2)\right\}. \tag{3.6.7}$$

The relations (3.6.4) and (3.6.5) yield

$$\|\mathbf{r}_{\varepsilon i}\|_{L^2(G)} = O(\sqrt{\varepsilon}). \tag{3.6.8}$$

Let us subtract equation (3.5.1) from (3.6.6). Then, for the difference

$$\mathbf{w}(x) = \tilde{\mathbf{u}}_\varepsilon^{(2)}(x) - \mathbf{u}_\varepsilon(x),$$

we get the problem

$$\begin{cases} L_\varepsilon \mathbf{w} = \mathbf{r}_\varepsilon(x) - \sum_{i=1}^d \frac{\partial}{\partial x_i} \mathbf{r}_{\varepsilon i}(x), & x \in G, \\ \mathbf{w}|_{\partial G} = 0. \end{cases} \tag{3.6.9}$$

Applying the *a priori* estimate (3.4.10) of Chapter 2, we get

$$\|\mathbf{w}\|_{H^1(G)} \le C_D' \left\{ C_{PF}\|\mathbf{r}_\varepsilon\|_{L^2(G)} + \sqrt{\sum_{i=1}^d \|\mathbf{r}_{\varepsilon i}\|^2_{L^2(g)}} \right\}, \tag{3.6.10}$$

where C_D', C_{PF} are independent of ε. So,

$$\|\tilde{\mathbf{u}}_\varepsilon^{(2)} - \mathbf{u}_\varepsilon\|_{H^1(G)} = O(\sqrt{\varepsilon}).$$

Again taking into consideration (3.6.4) and (3.6.5), we see that

$$\|(1 - \chi_\varepsilon)(\varepsilon \mathbf{u}_1 + \varepsilon^2 \mathbf{u}_2)\|_{H^1(G)} = O(\sqrt{\varepsilon}),$$

and

$$\|\varepsilon^2 \mathbf{u}_2\|_{H^1(G)} = O(\varepsilon);$$

we obtain the main estimate for the error

$$\|\mathbf{u}_\varepsilon - (\mathbf{u}_0 + \varepsilon \mathbf{u}_1)\|_{H^1(G)} = O(\sqrt{\varepsilon}). \tag{3.6.11}$$

On the other hand,

$$\|\varepsilon \mathbf{u}_1\|_{L^2(G)} = O(\varepsilon),$$

and so

$$\|\mathbf{u}_\varepsilon - \mathbf{u}_0\|_{L^2(G)} = O(\sqrt{\varepsilon}). \tag{3.6.12}$$

The estimates (3.6.11) and (3.6.12) justify the asymptotic approximation and the homogenized problem.

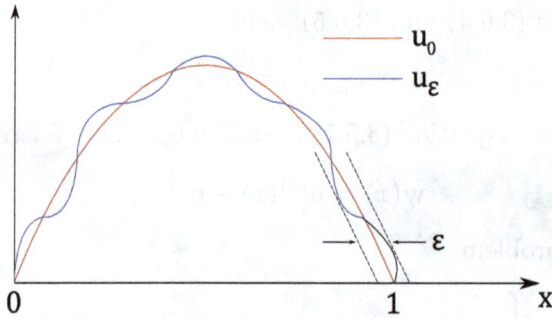

7. When the Equivalent Homogeneity Hypothesis is Wrong: Some Remarks on High-Contrast Media Homogenization

7.1. *Composite reinforced by highly conductive fibers*

Now, we discuss a special case of a composite material such that the equivalent homogeneity hypothesis is not true. In the above example, the structure of the solution is as follows: function u_0 independent of small parameter ε plus small (of orders ε and higher) rapidly oscillating correctors εu_1 and $\varepsilon^2 u_2$. We give an example in which the fluctuations of the solution are great (of order 1), and so any solution for a problem with constant coefficients is not close to the strongly fluctuating exact solution to the problem in the heterogeneous medium. This example is concerned with the second small parameter in the problem, i.e. the ratio of conductivities of compounds constituting the composite material. Consider three non-intersecting (and not touching one another) 1-periodic systems of parallel infinite cylinders oriented in the directions of axes ξ_1, ξ_2, and ξ_3 (we describe the geometry of inclusions in terms of fast variables ξ, see figure for the fiber-reinforcing composite material at the beginning of this chapter). Denote the space occupied by these systems of cylinders as B_1, B_2, and B_3, respectively. Define function $K_\omega(\xi)$ to be equal to some positive great number $\omega \gg 1$ for $\xi \in B_1 \cup B_2 \cup B_3$ and equal to 1 out of the set $B_1 \cup B_2 \cup B_3$ (the so-called high-contrast coefficient).

Consider a boundary value problem for the conductivity steady-state equation

$$-\mathrm{div}\left(K_\omega\left(\frac{x}{\varepsilon}\right)\nabla u_{\varepsilon\omega}\right) = 0, \quad x = (x_1, x_2, x_3) \in \Omega, \qquad (3.7.1)$$

set in a unit ball $\Omega \subset \mathbb{R}^3$ with the boundary condition at the unit sphere $\partial\Omega$:

$$u_{\varepsilon\omega} = \phi(x), \qquad (3.7.2)$$

where ϕ is the given smooth function.

The crucial condition for the validity of the equivalent homogeneity hypothesis is that $\varepsilon^2\omega \ll 1$. In this case, it is proved [15] that the solution to problem (3.7.1),(3.7.2) is close to the solution to the problem

$$-\mathrm{div}(\hat{K}\nabla u_0) = 0, \quad x = (x_1, x_2, x_3) \in \Omega, \qquad (3.7.3)$$

$$u_0|_{\partial\Omega} = \phi(x), \qquad (3.7.4)$$

where $\hat{K}$ is a diagonal matrix with the concentrations of the fibers of each coordinate direction on the diagonal, i.e. $\hat{K}_{ii} = \theta_i = mes(B_i \cap Q_1)$, where Q_1 is a unit cube. This means that

$$\max_{x\in\bar{\Omega}} |u_{\varepsilon\omega}(x) - u_0(x)| = O(\varepsilon\sqrt{\omega} + \omega^{-1/2}).$$

In the cases where $\varepsilon^2\omega \gg 1$ and $\varepsilon^2\omega = const$, one can find a function ϕ such that the equivalent homogeneity hypothesis is wrong. It can be proved that in the system of cylinders $B_{\varepsilon i} = \{x : x/\varepsilon \in B_i\}$ obtained by the homothetic contraction of B_i, with the factor $1/\varepsilon$ times, the following estimate holds:

$$\max_{x\in\bar{B}_{\varepsilon i}} |u_{\varepsilon\omega}(x) - u_{0i}(x)| = O((\varepsilon\sqrt{\omega})^{-1} + \varepsilon)$$

for $\varepsilon^2\omega \gg 1$ and

$$\max_{x\in\bar{B}_{\varepsilon i}} |u_{\varepsilon\omega}(x) - u_{0i}(x)| = O(\varepsilon)$$

for $\varepsilon^2\omega = const = \kappa$.

Here, the vector $(u_{01}(x), u_{02}(x), u_{03}(x))$ is a solution to the following multicomponent homogenized system of equations:

$$-\frac{\partial^2 u_{0i}}{\partial x_i^2} = 0, \quad i = 1, 2, 3, \quad x = (x_1, x_2, x_3) \in \Omega, \qquad (3.7.5)$$

$$u_{0i}|_{\partial\Omega} = \phi(x), \quad i = 1, 2, 3, \qquad (3.7.6)$$

for $\varepsilon^2 \omega \gg 1$ and respectively

$$-\frac{\partial^2 u_{0i}}{\partial x_i^2} + \kappa^{-1} \sum_{j=1}^{3} \hat{D}_{ij} u_{0j} = 0, \quad i = 1, 2, 3, \quad x = (x_1, x_2, x_3) \in \Omega,$$

$$(3.7.7)$$

$$u_{0i}|_{\partial\Omega} = \phi(x). \quad i = 1, 2, 3, \qquad (3.7.8)$$

for $\varepsilon^2 \omega = \mathrm{const} = \kappa$.

Here, $\hat{D}_{ij}$ are some constants; the algorithm of their computation is given in Refs. [15] and [16].

Functions u_{0i} may be very different, and so they provide great oscillations of order 1 at the characteristic distances between the fibers, which are of order ε. Physically, it means that the very high conductivity of the fibers translate the temperature from the boundary to inside the ball, and so some neighboring inner point fibers may have very different temperature if their ends at the boundary have very different temperatures.

7.2. *High-contrast spectral problems*

The previous section gives an example of a high-contrast media homogenization which is different from the classical homogenization in the case where $\varepsilon^2 \omega = \mathrm{const}$ or $\varepsilon^2 \omega \gg 1$. This example presents only one of multiple applications of high-contrast composite materials.

In particular, such materials exhibit the existence of the so-called band gaps, i.e. some intervals of the frequencies when the waves are stopped in the material and do not propagate through. These intervals depend on the microstructure of the material so that such materials can "absorb" the waves of given frequencies. The mathematical description of such materials is related to the spectral analysis of the operators considered previously in this chapter.

However, spectral problems are outside the scope of this book. So, we refer to an excellent presentation of this topic in Ref. [10].

Also the so-called dynamic materials with time-dependent macroscopic properties are studied in [25].

References

[1] I. Babuska. Solutions of interface problems by homogenization. *SIAM J. Math. Anal.*, Part 1, 7(5): 603–634; Part 2, 7(5): 635–645; Part 3, 8(6): 923–937, 1976.

[2] N.S. Bakhvalov. Averaged characteristics of media with periodic structure. *Dokl. Acad. Nauk SSSR*, 218: 1046–1048, 1974.

[3] N.S. Bakhvalov and G.P. Panasenko. *Homogenization: Averaging Processes in Periodic Media*. Moscow: Nauka; 1984; English translation: Dordrecht etc.: Kluwer, 1989.

[4] A. Bensoussan, J.L. Lions, and G. Papanicolaou. *Asymptotic Analysis for Periodic Structures*. Amsterdam: North-Holland, 1978.

[5] C. Conca. *Problèmes Mathématiques en Couplage Fluide-Structure*. Paris: Eyrolles, 1994.

[6] D. Cioranescu and P. Donato. *An Introduction to Homogenization*. Oxford: Oxford University Press, 1999.

[7] E. De Giorgi and S. Spagnolo. Sulla convergenza degli integrali dell'energia par operatori ellittici del secondo ordine. *Boll. Un. Mat. Ital.*, 8: 391–411, 1973.

[8] B. Delamotte. A hint of renormalization. (Lecture at the 9-th Vietnam School of Physics, Hué, January, 2003) *arXiv:hep-th/0212049* v3, 2003.

[9] V.V. Jikov, S.M. Kozlov, and O.A. Oleinik. *Homogenization of Partial Differential Operators and Integral Functionals*. Berlin-New York: Springer-Verlag, 1994. Russian version: *Homogenization of Differential Operators*. Moscow: Nauka (Fizmatlit), 1993.

[10] I.V. Kamotski and V.P. Smyshlyaev. Localized modes due to defects in high contrast periodic media via two-scale homogenization. *J. Math. Sci.*, 232(3): 349–377, 2018.

[11] V.A. Marchenko and E. Ya. Khruslov. *Boundary Value Problems in Domains with Fine-Grained Boundary*. Kiev: Naukova Dumka, 1974 (in Russian).

[12] C.C. Mei and B. Vernescu. *Homogenization Methods for Multiscale Mechanics*. Singapore: World Scientific Publishing Co., 2010.

[13] F. Murat. Compacité par compensation. *Ann. Scuola Norm. Sup. Pisa*, 5: 489–507, 1978.

[14] O.A. Oleinik, A.S. Shamaev, and G.A. Yosif'yan. *Mathematical Problems in Elasticity and Homogenization*. Amsterdam: Elsevier, 1992. Russian version: *Mathematical Problems in the Theory of Strongly Heterogeneous Elastic Media*. Moscow: Moscow University Publishers, 1990.

[15] G.P. Panasenko. Homogenization of processes in strongly non homogeneous structures. *Dokl. Acad. Nauk SSSR*, 298: 76–79, 1988; English translation: *Soviet Phys. Dokl.*, 33, 1988.

[16] G.P. Panasenko. Multicomponent homogenization of processes in strongly nonhomogeneous structures. *Math. Sbornik*, 181(1): 134–142, 1990; English translation: *Math. USSR Sbornik*, 69(1): 143–153, 1991.

[17] G.P. Panasenko. *Multi-scale Modelling for Structures and Composites.* Dordrecht: Springer, 2005.

[18] A. Pankov. *G-Convergence and Homogenization of Nonlinear Partial Differential Operators.* Dordrecht–Boston–London: Kluwer, 1997.

[19] A.L. Piatnitski, G.A. Chechkin, and A.S. Shamaev. *Homogenization. Methods and Applications.* Ed. Tamara Rozhkovskaya, Novosibirsk, 2007.

[20] E. Sanchez-Palencia. Equations aux dérivées partielles. Solutions périodiques par rapport aux variables d'espace et applications. *Compt. Rend. Acad. Sci. Paris*, sér. A., 271: 1129–1132, 1970.

[21] L. Tartar. *Homogenization.* Paris: Cours Peccot, Collège de France, 1977.

[22] A.N. Tikhonov and A.A. Samarskii. *Equations of Mathematical Physics.* Moscow: Nauka, 1977; English translation: Dover, 1990.

[23] K.G. Wilson and J. Kogut. The renormalization group and the ϵ — expansion. *Phys. Rep. C*, 12: 75–199, 1974.

[24] V.B. Berdichevsky, On the averaging of periodic structures. *Dokl. Akad. Nauk SSSR*, 222: 565–567, 1975.

[25] K. Lurie, *An Introduction to the Mathematical Theory of Dynamic Materials*, Springer, 2017.

Chapter 4

Dimension Reduction and Multiscale Modeling for Thin Structures

We consider as thin structures cylinders or rectangles with the characteristic diameter of cross-section $l << L$, where L is the length of the cylinder or rectangle. Also, we consider finite unions of such cylinders or rectangles.

Such domains describe the geometry of industrial installations, such as Tour Eiffel or the geometry of network of blood vessels.

First, let us introduce the main ideas of the asymptotic approach of dimension reduction. To this end, we consider the simplest thin domain: a thin rectangle. We assume that the small parameter ε is the ratio of its thickness l and its length L, i.e. $\varepsilon = l/L << 1$.

1. Dimension Reduction for the Poisson Equation in a Thin Rectangle: The Case of the Neumann Boundary Condition at the Lateral Boundary

Consider the thin rectangle

$$G_\varepsilon = (0,1) \times (0,\varepsilon) \subset \mathbb{R}^2,$$

where ε is a small positive parameter. Denote Γ_0 and Γ_1 as the parts of ∂G_ε corresponding to $x_1 = 0$ and $x_1 = 1$, respectively.

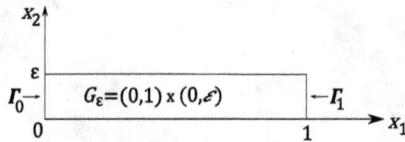

Consider the following problem:

$$\begin{cases} -\Delta u_\varepsilon = f(x_1), & x \in G_\varepsilon, \\ -\frac{\partial u_\varepsilon}{\partial n} = 0, & \partial G_\varepsilon \backslash (\Gamma_0 \cup \Gamma_1), \\ u_\varepsilon = \varphi\left(\frac{x_2}{\varepsilon}\right), & x \in \Gamma_0 = \{x_1 = 0\} \cap \partial G_\varepsilon, \\ u_\varepsilon = 0, & x \in \Gamma_1 = \{x_1 = 1\} \cap \partial G_\varepsilon, \end{cases} \quad (4.1.1)$$

where

$$f \in L^2((0,1)), \quad \varphi \in C^{(2)}([0,1]), \quad \varphi'(0) = \varphi'(1) = 0.$$

Let us apply the *dimensional reduction* method. Consider the following reduced one-dimensional (1-D) problem on $(0,1)$:

$$\begin{cases} -v''(x_1) = f(x_1), & x_1 \in (0,1), \\ v(0) = \langle \varphi \rangle, & v(1) = 0, \end{cases} \quad (4.1.2)$$

where $\langle \varphi \rangle = \int_0^1 \varphi(\xi)d\xi$.

We see that its solution $v(x_1)$ satisfies the equation and the boundary conditions at ∂G_ε everywhere except for Γ_0. In order to fix this mismatch, let us consider the following *boundary layer* problem in $\Pi = \mathbb{R}_+ \times (0,1)$:

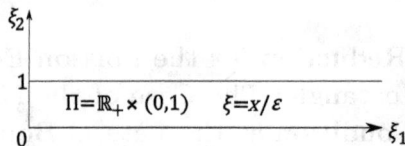

Find the $u_0^{BL}\left(\frac{x}{\varepsilon}\right)$ solution to

$$\begin{cases} -\Delta_\xi u_0^{BL} = 0, & \xi \in \Pi, \\ -\frac{\partial u_0^{BL}}{\partial n_\xi} = 0, & \xi_2 = 0 \text{ or } \xi_2 = 1, \\ u_0^{BL}(0, \xi_2) = \varphi(\xi_2) - \langle \varphi \rangle, \end{cases} \quad (4.1.3)$$

and the function $u_0^{BL}\left(\frac{x}{\varepsilon}\right)$ is called the boundary layer corrector. Let us prove that the bounded solution (4.1.3) exponentially decays as $\xi_1 \to +\infty$. To this end, expand $\varphi - \langle\varphi\rangle$ and u_0^{BL} in Fourier series with respect to ξ_2:

$$\varphi - \langle\varphi\rangle = \sum_{n=1}^{+\infty} a_n \cos(\pi\xi_2 n),$$

$$u_0^{BL}(\xi_1, \xi_2) = \sum_{n=1}^{+\infty} A_n(\xi_1) \cos(\pi\xi_2 n).$$

For A_n, we get the following ordinary differential equations (ODEs):

$$\begin{cases} -A_n'' + (\pi n)^2 A_n = 0, \quad \xi_1 > 0, \\ A_n(0) = a_n. \end{cases}$$

So,

$$u_0^{BL}(\xi) = \sum_{n=1}^{+\infty} a_n e^{-\pi n\xi_1} \cos(\pi\xi_2 n). \tag{4.1.4}$$

This result is called theorem of Phrägmen–Lindelöf type which states that

$$\exists C_1, C_2 > 0 : |u_0^{BL}|, |\nabla u_0^{BL}| \leq C_1 e^{-C_2\xi_1}. \tag{4.1.5}$$

An asymptotic solution is the sum

$$u_\varepsilon^a = v(x_1) + u_0^{BL}\left(\frac{x}{\varepsilon}\right), \tag{4.1.6}$$

and for any integer J, the following estimate holds:

$$\|u_\varepsilon^a - u_\varepsilon\|_{H^1(G_\varepsilon)} = O(\varepsilon^J). \tag{4.1.7}$$

The proof is given in the next section.

Let us find the distance from Γ_0 where the boundary layer corrector $u_0^{BL}\left(\frac{x}{\varepsilon}\right)$ has the "tail" smaller than ε^J. Using estimates (4.1.5), we get the following equation:

$$C_1 e^{-C_2 \frac{\delta}{\varepsilon}} < \varepsilon^J.$$

So, the thickness of the boundary layer area is

$$\delta \sim \text{const} \times J\varepsilon |\ln \varepsilon|.$$

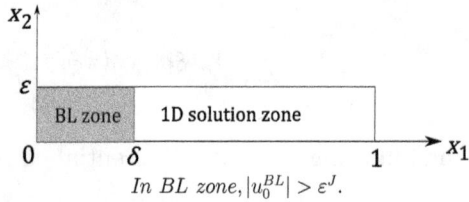

In BL zone, $|u_0^{BL}| > \varepsilon^J$.

2. Asymptotic Coupling of Models of Different Dimensions: Method of Asymptotic Partial Decomposition of the Domain (MAPDD)

In this section, we discuss the problem of coupling of models of different dimensions. Many real-world problems are related to solving partial differential equations in domains of complex geometry, combining multiple thin parts with massive parts, such as a set of blood vessels, structures in aircraft and spacecraft, industrial installations, and pipelines with reservoirs. Direct numerical computations with standard codes are impossible because such complex geometries need a very fine mesh "feeling" all elements of the structure, and so the three-dimensional (3-D) computations need too much time–memory resources. This is why the dimension reduction is a very popular trend in reducing computational cost; however, the completely reduced models loose very important local information and are not precise. For example, in blood circulation modeling, 1-D models are widely applied but the description of the clot formation and blood flow near a stent needs a 3-D local zoom. How to glue the models of different dimensions? We describe an asymptotic approach to this problem, based on the asymptotic analysis of partial differential equations in domains containing thin parts, connected sets of thin cylinders. For example, the Navier–Stokes equations are used in the hemodynamic modeling. We present the *method of partial asymptotic decomposition of domains (MAPDD)* which gives a high-precision coupling of models of different dimensions.

Consider the model problem (4.1.1) and reduce the equation in the part of the domain G_ε, where the boundary layer is less than ε^J, i.e. for $x_1 > \delta$, keeping the full dimension in the small part $G_{\varepsilon,\delta} = (0,\delta) \times (0,\varepsilon)$, i.e.

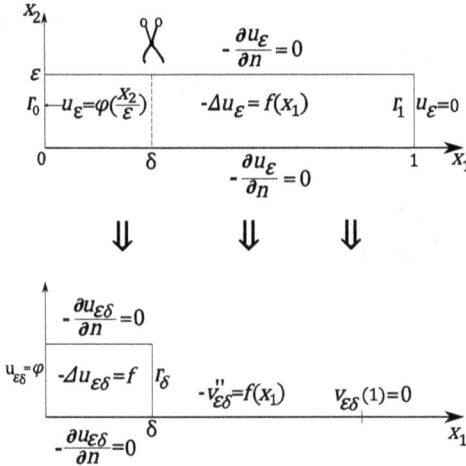

2-D/1-D junction conditions:

$$\begin{cases} u_{\varepsilon\delta}(\delta, x_2) = v_{\varepsilon\delta}(\delta), \\ \dfrac{1}{\varepsilon} \displaystyle\int_0^\varepsilon -\dfrac{\partial u_\varepsilon \delta}{\partial x_1}(\delta, x_2)dx_2 = -v'_{\varepsilon\delta}(\delta). \end{cases}$$

$$-\Delta u_{\varepsilon,\delta} = f(x_1), \quad x \in G_{\varepsilon,\delta},$$

$$-\frac{\partial u_{\varepsilon,\delta}}{\partial n} = 0, \quad x \in \partial G_{\varepsilon,\delta}\backslash(\Gamma_0 \cup \Gamma_\delta),$$

$$u_{\varepsilon,\delta} = \varphi\left(\frac{x_2}{\varepsilon}\right), \quad x \in \Gamma_0, \qquad (4.2.1)$$

$$-v''_{\varepsilon,\delta} = f(x_1), \quad x_1 \in (\delta, 1),$$

$$v_{\varepsilon,\delta}(1) = 0,$$

where $\Gamma_\delta = \{x_1 = \delta, x_2 \in (0,\varepsilon)\}$, and let us set at the interface Γ_δ the following conditions relating function $u_{\varepsilon,\delta}$ depending on (x_1, x_2)

to the function $v_{\varepsilon,\delta}$ depending on x_1 only:

$$u_{\varepsilon,\delta}(\delta, x_2) = v_{\varepsilon,\delta}(\delta),$$

$$\frac{1}{\varepsilon} \int_0^\varepsilon \left(-\frac{\partial u_{\varepsilon,\delta}}{\partial x_1}(\delta, v_2) \right) dx_2 = -v'_{\varepsilon,\delta}(\delta). \tag{4.2.2}$$

Problems (4.1.1) and (4.2.1), (4.2.2) are well posed (each of the two problems has a unique solution satisfying an *a priori* estimate), and extending $u_{\varepsilon,\delta}$ to the domain $G_\varepsilon \backslash G_{\varepsilon,\delta}$ as

$$u_{\varepsilon,\delta}(x_1, x_2) = v_{\varepsilon,\delta}(x_1),$$

one can prove the following.

Theorem 2.1. *Given* $J \in \mathbb{N}$, *there exists a constant* C *independent of* J *such that, if* $\delta = CJ\varepsilon|\ln \varepsilon|$, *then*

$$\|u_\varepsilon - u_{\varepsilon,\delta}\|_{H^1(G_\varepsilon)} = O(\varepsilon^J).$$

This theorem is proved in Ref. [1].

In order to prove this theorem, we multiply the boundary layer function by a cut-off function

$$\eta_\delta(x_1) = \eta \left(\frac{x_1}{\delta} \right),$$

where $\delta < 1/2$,

$$\eta(y) = \begin{cases} 1 \text{ for } y \in [0, 1], \\ 0 \text{ for } y \in [2, +\infty), \end{cases}$$

$\eta \in C^{(2)}([0, +\infty))$,

$$\tilde{u}^a_{\varepsilon,\delta} = v(x_1) + u_0^{BL} \left(\frac{x}{\varepsilon} \right) \eta \left(\frac{x_1}{\delta} \right),$$

and δ corresponds to ε^{J+2}, i.e. $C_1 e^{-C_2(\delta/\varepsilon)} < \varepsilon^{J+2}$ with C_1, C_2 constants from (4.1.5). Then, one can check that this function satisfies equation (4.1.1) with the discrepancy $O(\varepsilon^J)$

$$-\Delta \tilde{u}^a_{\varepsilon,\delta} = f(x_1) + r_{\varepsilon,\delta}(x), \tag{4.2.3}$$

and exactly the boundary conditions.

Here,

$$\|r_{\varepsilon,\delta}\|_{L^2(G_\varepsilon)} = O(\varepsilon^J). \tag{4.2.4}$$

Passing to the weak formulations for u_ε and $u_{\varepsilon,\delta}$, we introduce the functional

$$I(u, v) = \int_{G_\varepsilon} \nabla u \cdot \nabla v dx + \int_{G_\varepsilon} fv dx, \tag{4.2.5}$$

Sobolev space

$$H^1_{0\Gamma}(G_\varepsilon) = \{v \in H^1(G_\varepsilon) | v = 0 \text{ on } \Gamma_0 \cup \Gamma_1\},$$

subspace

$$H_{\varepsilon,\delta,\text{dec},0} = \left\{v \in H^1_{0\Gamma}(G_\varepsilon) \Big| \frac{\partial v}{\partial x_2} = 0 \text{ in } G_\varepsilon \backslash G_{\varepsilon,\delta}\right\},$$

and space

$$H_{\varepsilon,\delta,\text{dec}} = \left\{v \in H^1(G_\varepsilon) | v|_{\Gamma_1} = 0, \quad \frac{\partial v}{\partial x_2} = 0 \text{ in } G_\varepsilon \backslash G_{\varepsilon,\delta}\right\}.$$

Then, the weak formulation for (4.1.1) is as follows:
Find $u_\varepsilon \in H^1(G_\varepsilon)$ such that

$$u_\varepsilon(0, x_2) = \varphi\left(\frac{x_2}{\varepsilon}\right), \quad u_\varepsilon(1, x_2) = 0,$$

and $\forall v \in H^1_{0\Gamma}(G_\varepsilon)$,

$$I(u_\varepsilon, v) = 0; \tag{4.2.6}$$

for equation (4.2.3), with the boundary conditions of (4.1.1), we have:

$$-\text{find } \tilde{u}_{\varepsilon,\delta} \in H^1(G_\varepsilon),$$

such that

$$\tilde{u}_{\varepsilon\delta}(0, x_2) = \varphi\left(\frac{x_2}{\varepsilon}\right), \quad \tilde{u}_{\varepsilon\delta}(0, x_2) = 0,$$

and

$$\forall v \in H^1_{0\Gamma}(G_\varepsilon), \quad I(\tilde{u}_{\varepsilon,\delta}, v) = \int_{G_\varepsilon} r_{\varepsilon,\delta} v dx. \tag{4.2.7}$$

The weak formulation for the problem (4.2.1), (4.2.2) is as follows:

$$-\text{find } u_{\varepsilon,\delta} \in H_{\varepsilon,\delta,\text{dec}},$$

such that

$$u_{\varepsilon,\delta}(0, x_2) = \varphi\left(\frac{x_2}{\varepsilon}\right), \quad u_{\varepsilon,\delta}(1, x_2) = 0,$$

and

$$\forall v \in H_{\varepsilon,\delta,\text{dec},0}, \quad I(u_{\varepsilon,\delta}, v) = 0. \tag{4.2.8}$$

Subtracting (4.2.6) from (4.2.7) and (4.2.8) from (4.2.7), we get

$$\forall v \in H^1_{0\Gamma}(G_\varepsilon), \quad \int_{G_\varepsilon} \nabla(u_\varepsilon - \tilde{u}_{\varepsilon,\delta}) \cdot \nabla v \, dx = -\int_{G_\varepsilon} r_{\varepsilon,\delta} v \, dx, \tag{4.2.9}$$

and

$$\forall v \in H_{\varepsilon,\delta,\text{dec},0}, \quad \int_{G_\varepsilon} \nabla(u_{\varepsilon,\delta} - \tilde{u}_{\varepsilon,\delta}) \cdot \nabla v \, dx = -\int_{G_\varepsilon} r_{\varepsilon,\delta} v \, dx. \tag{4.2.10}$$

Taking $v = u_\varepsilon - \tilde{u}_{\varepsilon,\delta}$ in (4.2.9) and $u_{\varepsilon,\delta} - \tilde{u}_{\varepsilon,\delta}$ in (4.2.10) and applying then the CBS inequality and Poincaré–Friedrichs inequality, we get

$$\|\nabla(u_\varepsilon - \tilde{u}_{\varepsilon,\delta})\|_{L^2(G)} \le C\|r_{\varepsilon,\delta}\|_{L^2(G)},$$

$$\|\nabla(u_{\varepsilon,\delta} - \tilde{u}_{\varepsilon,\delta})\|_{L^2(G)} \le C\|r_{\varepsilon,\delta}\|_{L^2(G)}.$$

Applying the triangle inequality, we get

$$\|\nabla(u_\varepsilon - u_{\varepsilon,\delta})\|_{L^2(G_\varepsilon)} = O(\varepsilon^J),$$

and so

$$\|u_\varepsilon - u_{\varepsilon,\delta}\|_{H^1(G_\varepsilon)} = O(\varepsilon^J). \tag{4.2.11}$$

This estimate justifies the MAPDD in this model case.

Using the same scheme, one can justify this method for more complex thin rod structures which are the finite unions of thin cylinders (or rectangles in two-dimensional (2-D) case).

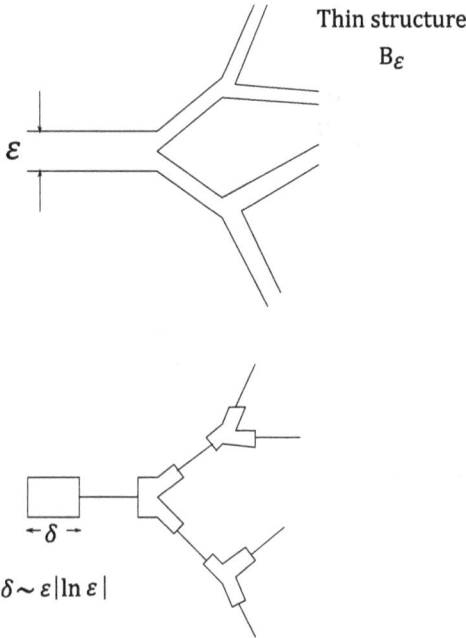

Thin structure
B_ε

ε

$\leftarrow \delta \rightarrow$

$\delta \sim \varepsilon |\ln \varepsilon|$

Partial asymptotic decomposition of the domain (partial asymptotic reduction).

3.   Dimension Reduction for the Poisson Equation in a Thin Rectangle: Case of the Dirichlet Boundary Condition

Consider the Dirichlet's problem for Laplacian in a thin rectangle:

$$G_\varepsilon = (0,1) \times (0,\varepsilon),$$

$$\begin{cases} -\Delta u_\varepsilon = f(x_1), & x \in G_\varepsilon, \\ u_\varepsilon|_{\partial G_\varepsilon} = 0, \end{cases} \tag{4.3.1}$$

where $f \in C^{(\infty)}([0,1])$.

(1a) Case of $f \in C_0^{(\infty)}([0,1])$:

Let us seek an asymptotic solution in the form

$$u_\varepsilon^{(J)} = \sum_{l=0}^{J} \varepsilon^{l+2} u_l\left(x_1, \frac{x_2}{\varepsilon}\right), \tag{4.3.2}$$

where $u_l \in C^{(2)}([0,1]^2)$ and J is an even integer, $J \geq 0$.

Plugging $u_\varepsilon^{(J)}$ in (4.3.1), we obtain

$$-\Delta u_\varepsilon^{(J)} = -\frac{\partial^2}{\partial x_1^2} u_\varepsilon^{(J)} - \frac{\partial^2}{\partial x_2^2} u_\varepsilon^{(J)}$$

$$= \left\{ -\sum_{l=0}^{J} \varepsilon^{l+2} \frac{\partial^2 u_l}{\partial x_1^2}(x_1,\xi_2) - \sum_{l=0}^{J} \varepsilon^{l+2}\varepsilon^{-2} \frac{\partial^2 u_l}{\partial \xi_2^2}(x_1,\xi_2) \right\} \Big|_{\xi_2=\frac{x_2}{\varepsilon}}$$

$$= \left\{ -\sum_{l=0}^{J} \varepsilon^{l} \frac{\partial^2 u_l}{\partial \xi_2^2}(x_1,\xi_2) - \sum_{l'=2}^{J+2} \varepsilon^{l'} \frac{\partial^2 u_{l'-2}}{\partial x_1^2}(x_1,\xi_2) \right\} \Big|_{\xi_2=\frac{x_2}{\varepsilon},\, l'=l+2}$$

$$= -\sum_{l=0}^{J} \varepsilon^{l} \left(\frac{\partial^2 u_l}{\partial \xi_2^2} + \frac{\partial^2 u_{l-2}}{\partial x_1^2} \right) - \varepsilon^{J+1} \frac{\partial^2 u_{J-1}}{\partial x_1^2} - \varepsilon^{J+2} \frac{\partial^2 u_J}{\partial x_1^2}.$$

Here, $u_l = 0$ if $l < 0$.

Equating this expansion to the right-hand side $f(x_1)$, we get a chain of problems for u_l:

$$(4.3.3) \quad \begin{cases} -\frac{\partial^2}{\partial \xi_2^2} u_l(x_1,\xi_2) = \frac{\partial^2}{\partial x_1^2} u_{l-2}(x_1,\xi_2) + f\delta_{l0}, & \xi_2 \in (0,1), \\ u_l|_{\xi_2=0,1} = 0. \end{cases}$$

For $l = 0$,

$$u_0 = \frac{1}{2}x_2(x_2 - 1)(-f(x_1)).$$

One can check directly: for odd l, $u_l = 0$, and

$$(4.3.4) \quad \begin{cases} -\Delta u_\varepsilon^{(J)} = f(x_1) + r_\varepsilon(x), & x \in G_\varepsilon, \\ u_\varepsilon^{(J)}|_{\partial G_\varepsilon} = 0, \end{cases}$$

$$r_\varepsilon = \varepsilon^{J+2} \frac{\partial^2 u_J}{\partial x_1^2} = O(\varepsilon^{J+2}) \text{ in } L^2(G_\varepsilon)$$

(more exactly, $O(\varepsilon^{J+2}\sqrt{\text{mes } G_\varepsilon})$), and so taking into consideration that $u_\varepsilon^{(J)}$ vanishes for $x_1 = 0$ and $x_1 = 1$ due to the condition $f \in C_0^{(\infty)}([0,1])$ and using the *a priori* estimate, we get

$$\|u_\varepsilon - u_\varepsilon^{(J)}\|_{H^1(G_\varepsilon)} = O\left(\varepsilon^{J+2}\sqrt{\text{mes} G_\varepsilon}\right). \tag{4.3.5}$$

(1b) Case of $f = 0$ in some neighborhood of $x_1 = 1$ but $f \neq 0$ if $x_1 = 0$: In this case, the boundary condition at Γ_0 fails. So, let us introduce the boundary layer corrector

$$u_\varepsilon^{BL(J)} = \sum_{l=0}^{J} \varepsilon^{l+2} u_l^{BL}\left(\frac{x_1}{\varepsilon}, \frac{x_2}{\varepsilon}\right), \tag{4.3.6}$$

where $u_l^{BL}|_{x_1=0}$ compensates the trace $u_l|_{x_1=0}$:

$$u_l^{BL}|_{x_1=0} = -u_l|_{x_1=0}.$$

So, we get for u_l^{BL}, a chain of problems in dilated variables ξ belonging to the half-strip $\Pi = \mathbb{R}_+ \times (0,1)$:

$$\begin{cases} -\Delta_\xi u_l^{BL} = 0, & \xi \in \Pi, \\ u_l^{BL} = 0, & \text{if } \xi_2 = 0, 1, \\ u_l^{BL}(0, \xi_2) = -u_l(0, \xi_2). \end{cases} \tag{4.3.7}$$

As in Section 1, let us prove that the solutions of (4.3.7) decay exponentially as $\xi_1 \to +\infty$. To this end, we use the Fourier series: let

$$-u_l(0, \xi_2) = \sum_{n=1}^{\infty} b_n \sin(\xi_2 \pi n),$$

then we seek

$$u_l^{BL}(\xi_1, \xi_2) = \sum_{n=1}^{\infty} B_n(\xi_1) \sin(\xi_2 \pi n),$$

and for B_n, we get the ODEs

$$\begin{cases} -B_n'' + (\pi n)^2 B_n = 0, & \xi_1 \in \mathbb{R}_+, \\ B_n(0) = b_n, \\ B_n(\xi_1) \to 0 \text{ as } \xi_1 \to +\infty. \end{cases}$$

So, $B_n = b_n e^{-\pi n \xi_1}$,

$$u_l^{BL} = \sum_{n=1}^{\infty} b_n e^{-\pi n \xi_1} \sin(\xi_2 \pi n). \qquad (4.3.8)$$

So, we prove the theorem of Phrägmen–Lindelöf type:

$$\exists C_1, C_2 > 0 : \left|u_l^{BL}\right|, \left|\nabla u_l^{BL}\right| \le C_1 e^{-C_2 \xi_1}, (C_2 = \pi).$$

In order to obtain the error estimate, we introduce the "corrected" asymptotic solution:

$$u_\varepsilon^{(J)}\left(x_1, \frac{x_2}{\varepsilon}\right) + u_\varepsilon^{BL(J)}\left(\frac{x}{\varepsilon}\right) \eta(3x_1),$$

where

$$\eta(t) = \begin{cases} 1, & \text{if } |t| \le 1, \\ 0, & \text{if } |t| \ge 2, \end{cases}$$

$\eta \in C^{(2)}(\mathbb{R})$ is a cut-off function.

Calculating the residual, we get

$$r_\varepsilon - \left(\Delta u_\varepsilon^{BL(J)} \eta(3x_1) + 2\frac{\partial u_\varepsilon^{BL(J)}}{\partial x_1} \eta'(3x_1) + u_\varepsilon^{BL(J)} \eta''(3x_1)\right)$$

$$= O\left(\varepsilon^{J+2} \sqrt{\text{mes } B_\varepsilon}\right).$$

So, we get an error estimate

$$\left\| u_\varepsilon - \left(u_\varepsilon^{(J)} + u_\varepsilon^{BL(J)} \eta \right) \right\|_{H^1} = O \left(\varepsilon^{J+2} \sqrt{\mathrm{mes} B_\varepsilon} \right).$$

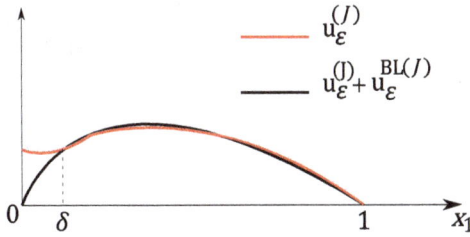

As in Section 1, we evaluate the thickness δ of the boundary layer. Using the estimate

$$\left| u_\varepsilon^{BL(J)}(x) \right| \le C_1 e^{-C_2 x_1 / \varepsilon},$$

we see that it is smaller than ε^J for $x_1 \ge \delta$ such that

$$C_1 e^{-C_2 \delta / \varepsilon} \le \varepsilon^J,$$

and we obtain the following formula for δ:

$$\delta = \mathrm{const} J \varepsilon |\ln \varepsilon|.$$

4. Dirichlet's Problem for Laplacian in a Thin Tube Structure

Consider the domain $B_\varepsilon = B_\varepsilon^1 \cup B_\varepsilon^2$, where $B_\varepsilon^1 = (-1, 1) \times \left(-\frac{\varepsilon}{2}, \frac{\varepsilon}{2} \right)$, $B_\varepsilon^2 = \left(-\frac{\varepsilon}{2}, \frac{\varepsilon}{2} \right) \times (0, 1)$.

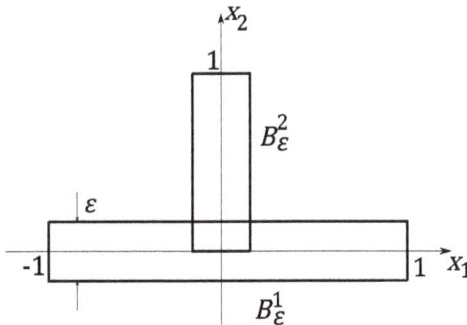

Consider the following problem:

$$\begin{cases} -\Delta u_\varepsilon = 1 \ \text{in} \ B_\varepsilon, \\ u_\varepsilon|_{\partial B_\varepsilon} = 0. \end{cases} \tag{4.4.1}$$

Let us describe the construction of an asymptotic expansion.
Step 1. Define "Poiseuille"-type solution:

$$U_\varepsilon^1(x) = -\tfrac{1}{2}\left(x_2^2 - \tfrac{\varepsilon^2}{4}\right) \ \text{in} \ B_\varepsilon^1,$$

$$U_\varepsilon^2(x) = -\tfrac{1}{2}\left(x_1^2 - \tfrac{\varepsilon^2}{4}\right) \ \text{in} \ B_\varepsilon^2. \tag{4.4.2}$$

Step 2. Multiply U_ε^1 and U_ε^2 by the cut-off functions

$$u_{12}^\varepsilon(x) = U_\varepsilon^1(x)\left(1 - \eta\left(\tfrac{x_1}{3\varepsilon}\right)\right) + U_\varepsilon^2(x)\left(1 - \eta\left(\tfrac{x_2}{3\varepsilon}\right)\right). \tag{4.4.3}$$

Plugging it in the equation, we get

$$-\Delta u_{12}^\varepsilon = 1 + F\left(\tfrac{x}{\varepsilon}\right),$$

$$F(\xi) = 1 - \eta\left(\tfrac{\xi_1}{3}\right) - \tfrac{1}{18}\left(\xi_2^2 - \tfrac{1}{4}\right)\eta''\left(\tfrac{\xi_1}{3}\right)$$

$$+ 1 - \eta\left(\tfrac{\xi_2}{3}\right) - \tfrac{1}{18}\left(\xi_1^2 - \tfrac{1}{4}\right)\eta''\left(\tfrac{\xi_2}{3}\right)$$

$$- 1 \ \text{for} \ \xi \in (-6,6)^2 \cap \Pi,$$

$$\Pi = \mathbb{R} \times \left(-\tfrac{1}{2},\tfrac{1}{2}\right) \cup \left(\tfrac{1}{2},\tfrac{1}{2}\right) \times \mathbb{R}_+;$$

$F = 0$ out of $(-6,6)^2$.
Step 3. To compensate residual F introduce $u_{BL}(\xi)$, solution to

$$\begin{cases} -\Delta_\xi u_{BL} = -F(\xi), \quad \xi \in \Pi, \\ u_{BL}|_{\partial \Pi} = 0. \end{cases}$$

Clearly, there exists a unique solution from $H_0^1(\Pi)$ and $|u_{BL}|, |\nabla u_{BL}| \leq C_1 e^{-C_2|\xi|}$.

Step 4. Construct boundary layer correctors (as before) to compensate traces of u_{12}^ε on

$$\gamma^{(-1,0)} = \left\{ x_1 = -1, x_2 \in \left(-\frac{\varepsilon}{2}, \frac{\varepsilon}{2} \right) \right\} : U_\varepsilon^{BL(-1,0)}(\xi),$$

$$\gamma^{(1,0)} = \left\{ x_1 = 1, x_2 \in \left(-\frac{\varepsilon}{2}, \frac{\varepsilon}{2} \right) \right\} : U_\varepsilon^{BL(1,0)}(\xi),$$

$$\gamma^{(0,1)} = \left\{ x_1 \in \left(-\frac{\varepsilon}{2}, \frac{\varepsilon}{2} \right), x_2 = 1 \right\} : U_\varepsilon^{BL(0,1)}(\xi).$$

Finally, the asymptotic approximation is

$$u_\varepsilon^a = U_\varepsilon^1(x) \left(1 - \eta \left(\frac{x_1}{3\varepsilon} \right) \right) + U_\varepsilon^2(x) \left(1 - \eta \left(\frac{x_2}{3\varepsilon} \right) \right)$$

$$+ U_{BL} \left(\frac{x}{\varepsilon} \right) \eta(3x_1)\eta(3x_2)$$

$$+ U_\varepsilon^{BL(-1,0)} \left(\frac{x_1+1}{\varepsilon}, \frac{x_2}{\varepsilon} \right) \eta(3(x_1+1))$$

$$+ U_\varepsilon^{BL(1,0)} \left(\frac{x_1-1}{\varepsilon}, \frac{x_2}{\varepsilon} \right) \eta(3(x_1-1))$$

$$+ U_\varepsilon^{BL(0,1)} \left(\frac{x_1}{\varepsilon} \frac{x_2-1}{\varepsilon} \right) \eta(3(x_2-1)), \quad \varepsilon < 1/9.$$

All boundary layers decay exponentially by Phrägmen–Lindelöf theorems, proved by Fourier analysis.

Calculating the residual, we see that it is of order $O(\varepsilon^{J+2}\sqrt{\text{mes } B_\varepsilon})$ in L^2 and applying an *a priori* estimate we get as before

$$\|u_\varepsilon^a - u_\varepsilon\|_{H^1(B_\varepsilon)} = O(\varepsilon^{J+2}\sqrt{\text{mes } B_\varepsilon}).$$

5. Method of Asymptotic Partial Decomposition of Domain for a T-shaped Domain

Decompose

$$B_\varepsilon = B_{\varepsilon\delta}^1 \cup B_{\varepsilon\delta}^2 \cup B_{\varepsilon\delta}^{(0,0)} \cup B_{\varepsilon\delta}^{(-1,0)} \cup B_{\varepsilon\delta}^{(1,0)} \cup B_{\varepsilon\delta}^{(0,1)},$$

where

$$B_{\varepsilon\delta}^1 = (-1+\delta, -\delta) \times \left(-\frac{\varepsilon}{2}, \frac{\varepsilon}{2}\right) \cup (\delta, 1-\delta) \times \left(-\frac{\varepsilon}{2}, \frac{\varepsilon}{2}\right),$$

$$B_{\varepsilon\delta}^2 = \left(-\frac{\varepsilon}{2}, \frac{\varepsilon}{2}\right) \times (\delta, 1-\delta),$$

$$B_{\varepsilon\delta}^{(0,0)} = (-\delta, \delta) \times \left(-\frac{\varepsilon}{2}, \frac{\varepsilon}{2}\right) \cup \left(-\frac{\varepsilon}{2}, \frac{\varepsilon}{2}\right) \times (0, \delta),$$

$$B_{\varepsilon\delta}^{(-1,0)} = (-1, -1+\delta) \times \left(-\frac{\varepsilon}{2}, \frac{\varepsilon}{2}\right),$$

$$B_{\varepsilon\delta}^{(1,0)} = (1-\delta, 1) \times \left(-\frac{\varepsilon}{2}, \frac{\varepsilon}{2}\right),$$

$$B_{\varepsilon\delta}^{(0,1)} = \left(-\frac{\varepsilon}{2}, \frac{\varepsilon}{2}\right) \times (1-\delta, 1).$$

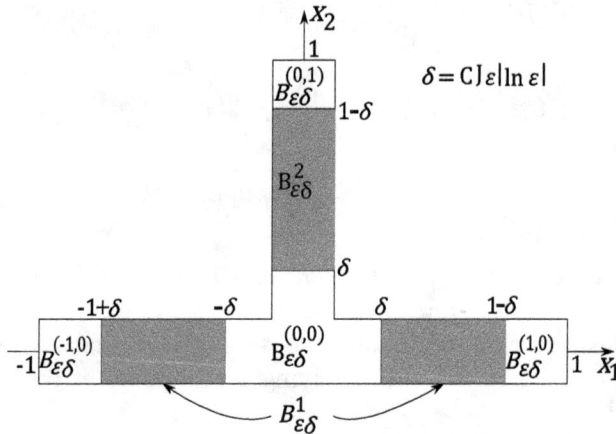

Consider an approximation u_d: it is the solution to equation $-\Delta u_d = 1$;

in $B_{\varepsilon\delta}^{(0,0)}$ with boundary condition $u_d = 0$ on $\partial B_\varepsilon \cap \partial B_{\varepsilon\delta}^{(0,0)}$, $u_d = U_\varepsilon^1$ on $\partial B_{\varepsilon\delta}^{(0,0)} \cap \{x_1 = \pm\delta\}$, $u_d = U_\varepsilon^2$ on $\partial B_{\varepsilon\delta}^{(0,0)} \cap \{x_2 = \delta\}$;

in $B_{\varepsilon\delta}^{(\pm 1,0)}$ with boundary condition $u_d = 0$ on $\partial B_\varepsilon \cap \partial B_{\varepsilon\delta}^{(+1,0)}$, $u_d = U_\varepsilon^1$ on $\partial B_{\varepsilon\delta}^{(\pm 1,0)} \cap \{x_1 = \pm(1-\delta)\}$;

in $B_{\varepsilon\delta}^{(0,1)}$ with boundary condition $u_d = 0$ on $\partial B_\varepsilon \cap \partial B_{\varepsilon\delta}^{(0,1)}$, $u_d = U_\varepsilon^2$ on $\{x_2 = 1-\delta\}$.

We define $u_d = U_\varepsilon^1$ in $B_{\varepsilon\delta}^1$, $u_d = U_\varepsilon^2$ in $B_{\varepsilon\delta}^2$.

The error estimate is given as follows.

Theorem 5.1.

$$\|u_\varepsilon - u_d\|_{H^1(B_\varepsilon)} = O\left(\varepsilon^{J+2}\sqrt{\text{mes } B_\varepsilon}\right).$$

This method economizes computational resources. The problem is reduced to independent problems in $B_{\varepsilon\delta}^{(0,0)}$, $B_{\varepsilon\delta}^{(\pm 1,0)}$, and $B_{\varepsilon\delta}^{(0,1)}$, which can be solved in parallel.

This approach is generalized in the case of the Stokes and Navier–Stokes equations in thin tube structures modeling the blood flow in a network of vessels [1,2].

6. Method of Asymptotic Partial Decomposition of Domain for Flows in a Tube Structure (Applications in Hemodynamics)

Being motivated by the modeling of blood flow in a network of blood vessels, we consider the Stokes equation in a tube structure. These domains are connected finite unions of thin finite cylinders (in the 2-D case, thin rectangles). Each such tube structure may be schematically represented by its graph: letting the thickness of tubes to zero, we find that tubes degenerate to segments.

6.1. Tube structure: Graphs

Definition 6.1. Let $O_1, O_2, \ldots, O_N$ be N different points in $\mathbb{R}^d$, $d = 2, 3$, and $e_1, e_2, \ldots, e_M$ be M closed segments each connecting two of these points (i.e. each $e_j = \overline{O_{i_j} O_{k_j}}$, where $i_j, k_j \in \{1, \ldots, N\}, i_j \neq k_j$). All points O_i are supposed to be the ends of some segments e_j. The segments e_j are called edges of the graph. A point O_i is called a node if it is the common end of at least two edges, and O_i is called a vertex if it is the end of the only one edge. Any two edges e_j and e_i can intersect only at the common node. The set of vertices is supposed to be non-empty.

Denote graph $\mathcal{B} = \bigcup_{j=1}^{M} e_j$ as the union of edges and assume that $\mathcal{B}$ is a connected set.

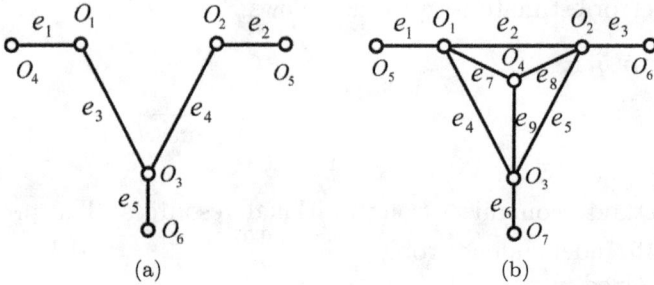

(a)   (b)

Let e be some edge, $e = \overline{O_i O_j}$. Consider two Cartesian coordinate systems in $\mathbb{R}^d$. The first one has the origin in O_i and the axis $O_i x_1^{(e)}$ has the direction of the ray $[O_i O_j)$; the second one has the origin in O_j and the opposite direction, i.e. $O_j \tilde{x}_1^{(e)}$ is directed over the ray $[O_j O_i)$.

Further, we choose one or another coordinate system denoting the local variable in both cases as x^e and pointing out which end is taken as the origin of the coordinate system.

With every edge e_j, we associate a bounded domain $\sigma_j \subset \mathbb{R}^{d-1}$ having a Lipschitz boundary $\partial \sigma_j$, $j = 1, \ldots, M$. For every edge $e_j = e$ and associated $\sigma_j = \sigma^{(e)}$, we denote by $\Pi_\varepsilon^{(e)}$ the cylinder

$$\Pi_\varepsilon^{(e)} = \left\{ x^{(e)} \in \mathbb{R}^d : x_1^{(e)} \in (0, |e|), \ \frac{x^{(e)\prime}}{\varepsilon} \in \sigma^{(e)} \right\},$$

where $x^{(e)\prime} = (x_2^{(e)}, \ldots, x_d^{(e)})$, $|e|$ is the length of the edge e, and $\varepsilon > 0$ is a small parameter. Note that the edges e_j and Cartesian coordinates of nodes and vertices O_j, as well as the domains σ_j, do not depend on ε.

Let $O_1, \ldots, O_{N_1}$ be nodes and $O_{N_1+1}, \ldots, O_N$ be vertices. Let $\omega^1, \ldots, \omega^{N_1}$ be bounded independent of ε domains in $\mathbb{R}^d$ with Lipschitz boundaries $\partial \omega^j$; introduce the nodal domains $\omega_\varepsilon^j = \{ x \in \mathbb{R}^d : \frac{x - O_j}{\varepsilon} \in \omega^j \}$.

Definition 6.2. By a *tube structure*, we refer to the following domain:

$$B_\varepsilon = \left(\bigcup_{j=1}^{M} \Pi_\varepsilon^{(e_j)} \right) \bigcup \left(\bigcup_{j=1}^{N_1} \omega_\varepsilon^j \right).$$

Suppose that it is a connected set and that the boundary ∂B_ε is C^2-regular except for the corners between the lateral surfaces of the cylinders $\Pi_\varepsilon^{(e_{N_1}+1)}, \ldots, \Pi_\varepsilon^{(e_N)}$ and their bases $\sigma_\varepsilon^{N_1+1}, \ldots, \sigma_\varepsilon^N$.

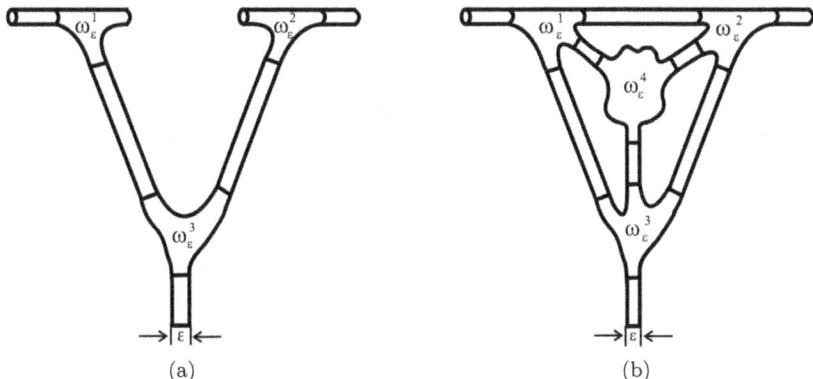

(a)   (b)

Tube structures.

6.2.   *Formulation of the problem*

Consider the Stokes equation in the tube structure B_ε:

$$-\nu\Delta\mathbf{v} + \nabla p = 0,$$
$$\operatorname{div}\mathbf{v} = 0, \qquad\qquad (4.6.1)$$
$$\mathbf{v}\big|_{\partial B_\varepsilon} = \mathbf{g}.$$

Assume that the fluid velocity $\mathbf{g}$ at the boundary ∂B_ε has the following structure: $\mathbf{g} = 0$ everywhere on ∂B_ε except for the set $\sigma_\varepsilon^{N_1+1}, \ldots, \sigma_\varepsilon^N$ of bases of the cylinders $\Pi_\varepsilon^{(e_{N_1}+1)}, \ldots, \Pi_\varepsilon^{(e_N)}$ associated to (i.e. containing) the vertices $O_{N_1+1}, \ldots, O_N$, i.e.

$$\mathbf{g}(x)\big|_{\sigma_\varepsilon^j} = \mathbf{g}^j\left(\frac{x-O_j}{\varepsilon}\right)\Big|_{\sigma_\varepsilon^j}, \quad j = N_1+1, \ldots, N,$$
$$\mathbf{g}(x)\Big|_{\partial B_\varepsilon\backslash\left(\bigcup_{j=N_1+1}^N \sigma_\varepsilon^j\right)} = 0, \qquad\qquad (4.6.2)$$

where $\mathbf{g}^j$ is a smooth function belonging to $C_0^{(2)}(\sigma_\varepsilon^j)$ such that $\operatorname{div}\mathbf{g}^j = 0$ on σ_ε^j, $\int_{\partial B_\varepsilon}\mathbf{g}\cdot\mathbf{n}ds = 0$, $\nu > 0$.

To define the weak solution, we use the approach of Section 7 in Chapter 2. Namely, it is a vector-valued function $\mathbf{v} \in H^1(B_\varepsilon)$

satisfying the conditions div $\mathbf{v} = 0$, $\mathbf{v}|_{\partial B_\varepsilon} = \mathbf{g}$ such that

$$\forall \mathbf{w} \in H_{\mathrm{div}} = \{\mathbf{u} \in H_0^1(B_\varepsilon)|\mathrm{div}\ \mathbf{u} = 0\},$$

$$\int_{B_\varepsilon} \nu\nabla\mathbf{v}\cdot\nabla\mathbf{w}dx = 0.$$

In the following, we describe the MAPDD for this problem; its justification is similar to the justification in the case of the Dirichlet boundary layer problem for the Poisson equation (Section 5).

6.3. *Partial asymptotic decomposition of the domain for the Stokes equation*

Let us define first the Poiseuille flow in an infinite tube $\Pi_\varepsilon = \mathbb{R} \times \sigma_\varepsilon^j$. It is an exact solution to the Stokes equations

$$-\nu\Delta\mathbf{v}_P + \nabla p_P = 0,$$

$$\mathrm{div}\ \mathbf{v}_P = 0,$$

$$\mathbf{v}_P|_{\partial\Pi_\varepsilon} = \mathbf{0},$$

and it has the form, in the case $d = 3$,

$$\mathbf{v}_P(x) = \begin{pmatrix} \alpha v_{P0}(x_2, x_3) \\ 0 \\ 0 \end{pmatrix},$$

$$p_P(x) = -\alpha x_1 + \beta,$$

where $\alpha, \beta \in \mathbb{R}$, $v_{P0}(x_2, x_3)$ satisfies

$$\begin{cases} -\nu\Delta v_{P0} = 1, (x_2, x_3) \in \sigma_\varepsilon^j, \\ v_{P0}|_{\partial\sigma_\varepsilon^j} = 0; \end{cases}$$

in the case $d = 2$, $\sigma_\varepsilon^j = \left(-\frac{\varepsilon}{2}, \frac{\varepsilon}{2}\right)$.

$$\mathbf{v}_P(x) = \begin{pmatrix} \alpha v_{P0}(x_2) \\ 0 \end{pmatrix},$$

$$p_P(x) = -\alpha x_1 + \beta, \quad v_P(x_2) = -\frac{1}{2\nu}\left(x_2^2 - \varepsilon^2/4\right).$$

Note that in the case where σ_ε^j is a disk of radius $\varepsilon/2$,

$$v_{P0}(x) = -\frac{1}{4\nu}\left(r^2 - \frac{\varepsilon^2}{4}\right).$$

Denote $(\tilde{\mathbf{v}}_P(x), \tilde{p}_P)$ as the Poiseuille flow in another coordinate system obtained by rotations, i.e. in the local coordinate system associated to e_j.

Let us describe the algorithm of the MAPDD for the Stokes problem set in a tube structure B_ε. Let δ be a small positive number much greater than ε (it is chosen to be of the order $O(\varepsilon|\ln \varepsilon|)$). For any edge $e = \overline{O_i O_j}$ of the graph, introduce two hyperplanes orthogonal to this edge and crossing it at the distance δ from its ends.

Denote the cross-sections of the cylinder $\Pi_\varepsilon^{(e)}$ by these two hyperplanes, respectively, by $S_{i,j}$ (the cross-section at the distance δ from O_i) and $S_{j,i}$ (the cross-section at the distance δ from O_j), and denote the part of the cylinder $\Pi_\varepsilon^{(e)}$ between these two cross-sections by $B_{ij}^{\mathrm{dec},\varepsilon}$. Let $B_i^{\varepsilon,\delta}$ be connected, truncated by the cross-sections $S_{i,j}$, part of B_ε which contains the vertex or the node O_i.

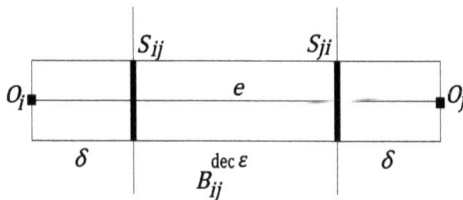

Truncation of the cylinder $\Pi_\varepsilon^{(e)}$.

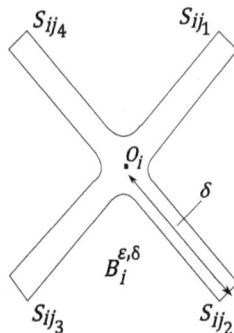

Connected component $B_i^{\varepsilon,\delta}$.

Introduce the space $H^1_{\text{div}=0}(B_\varepsilon)$ of all divergence-free vector-valued functions from the space $H^1(B_\varepsilon)$ vanishing for $x \in \partial B_\varepsilon \backslash (\cup_{j=N_1+1}^N \sigma_\varepsilon^j)$.

Define the subspace $H^1_{\text{div}=0}(B_\varepsilon, \delta)$ of $H^1_{\text{div}=0}(B_\varepsilon)$ such that on every truncated cylinder $B_{ij}^{\text{dec},\varepsilon}$, its elements (vector-valued functions) coincide with the Poiseuille type flows $\tilde{\mathbf{v}}_P$. We also consider the subspace $H^1_{0,\text{div}=0}(B_\varepsilon, \delta)$ of the space $H^1_{\text{div}=0}(B_\varepsilon, \delta)$ such that its elements vanish on the whole boundary ∂B_ε and the subspace $L^2(B_\varepsilon, \delta)$ of the space $L^2(B_\varepsilon)$ such that its elements are affine functions of $x_1^{(e)}$ on every truncated cylinder $B_{ij}^{\text{dec},\varepsilon}$.

The MAPDD replaces problem (4.6.1) by its projection on $H^1_{\text{div}=0}(B_\varepsilon, \delta)$: Find $\mathbf{v}_d \in H^1_{\text{div}=0}(B_\varepsilon, \delta)$, $\mathbf{v}_d|_{\partial B_\varepsilon} = \mathbf{g}$ such that $\forall \mathbf{w} \in H^1_{0,\text{div}=0}(B_\varepsilon, \delta)$,

$$\int_{B_\varepsilon} \nu \nabla \mathbf{v}_d \cdot \nabla \mathbf{w} dx = 0.$$

For numerical implementation, it is more convenient to use the weak formulation with pressure:

Find $\mathbf{v}_d \in H^1(B_\varepsilon, \delta), \mathbf{v}_d|_{\partial B_\varepsilon} = \mathbf{g}$, and $p_d \in L^2(B_\varepsilon, \delta)$ such that $\forall \mathbf{w} \in H^1_0(B_\varepsilon, \delta), q \in L^2(B_\varepsilon, \delta)$,

$$\int_{B_\varepsilon} \nu \nabla \mathbf{v}_d \cdot \nabla \mathbf{w} dx - \int_{B_\varepsilon} p_d \text{ div } \mathbf{w} dx + \int_{B_\varepsilon} q \text{ div } \mathbf{v}_d dx = 0,$$

where $H^1(B_\varepsilon, \delta)$, $H^1_0(B_\varepsilon, \delta)$ are the subspaces of the corresponding spaces with elements coinciding with the Poiseuille type functions $\tilde{\mathbf{v}}_P$ on $B_{ij}^{\text{dec},\varepsilon}$.

It is proved [2–4] that there exists a constant $C > 0$ such that given $J \in \mathbb{N}^*$ if $\delta = CJ\varepsilon|\ln \varepsilon|$, then the exact solution $(\mathbf{v}, p)$ and the MAPDD solution $(\mathbf{v}_d, p_d)$ satisfy the estimate

$$\|\mathbf{v}_d - \mathbf{v}\|_{H^1(B_\varepsilon)} + \|p_d - p\|_{L^2(B_\varepsilon)} = O(\varepsilon^J),$$

where p_d and p are normalized as follows:

$$\int_{B_\varepsilon} p_d dx = 0, \quad \int_{B_\varepsilon} p dx = 0.$$

This approach is justified for the non-stationary Navier–Stokes equations and non-Newtonian flows. Numerical experiments confirm the effectiveness and high precision of this method.

References

[1] G.P. Panasenko. *Multi-scale Modelling for Structures and Composites.* Dordrecht: Springer, 2005.
[2] G. Panasenko and K. Pileckas. Asymptotic analysis of the non-steady Navier-Stokes equations in a tube structure. I. The case without boundary layer-in-time. *Nonlinear Anal. Ser. A, Theory Methods Appl.*, 122: 125–168, 2015. Doi: 10.1016/j.na.2015.03.008.
[3] C. Bertoglio, C. Conca, D. Nolte, G. Panasenko, and K. Pileckas, Junction of models of different dimension for flows in tube structures by Womersley-type interface conditions, *SIAM J. Appl. Math.*, 79: 3, 959–985, 2019 https://doi.org/10.1137/M1229572.
[4] C. Bertoglio, D. Nolte, G. Panasenko, and K. Pileckas, Reconstruction of the pressure in the method of asymptotic partial decomposition for the flows in tube structures. *SIAM J. Appl. Math.*, 81: 5, 20832110, 2021. https://doi.org/10.1137/20M1388462.

Appendix A

Diffusion Equation with Dirac-like Potential: Model of a Periodic Set of Small Cells in a Nutrient

In Section 7 of Chapter 3, we have seen that in the case of presence in the model of multiple small parameters, the standard homogenization procedure may fail. So, in the case where the model contains two or more parameters, more detailed analysis is needed. This section gives an example of such an analysis in the case of the diffusion equation with periodic potential:

$$\frac{\partial u_\varepsilon}{\partial t} - \Delta u_\varepsilon + q(x/\varepsilon)u_\varepsilon = f(x,t),$$

where q is a periodic bounded measurable function.

Applying the method of Section 3.4, we can easily determine that the solution can be asymptotically approximated by the same equation with the averaged potential, i.e.

$$\frac{\partial u_0}{\partial t} - \Delta u_0 + \langle q \rangle u_0 = f(x,t).$$

However, if the support of the function q has a small measure but its average is constant, the answer is more complicated, and it depends on the relation between small parameters. The functions having support of the small measure and finite integral are generalizations of Dirac's function. To understand the difficulties of the formulation of partial differential equations with Dirac's potential, let us

125

consider the following "toy problem" having an analytical solution and introduced by V. Volpert.

1. On the Approximation of Dirac's Potential

Let us define a solution for elliptic equation with Dirac potential in the case of dimension $d > 1$. This problem is considered as a limit in some sense of the same equation with the potential vanishing everywhere except for a small disc (ball), where it is equal to a large parameter ω, so that its integral is equal to one.

Consider the equation

$$\Delta u - q(x)u = 0 \qquad (\text{A.1})$$

in the two-dimensional (2-D) disc $B = \{|x| \le 1\}$ with the boundary condition

$$u|_{\partial B} = 1. \qquad (\text{A.2})$$

Function q is taken in the form

$$q(x) = \begin{cases} \omega, & |x| < \delta, \\ 0, & |x| \ge \delta, \end{cases}$$

where $\omega > 1$ and $\delta < 1$ are positive constants related by the condition

$$\pi\delta^2 = \omega^{-1}.$$

This relation means that the integral of q is equal to one. If ω tends to $+\infty$, then q approximates Dirac's delta function.

Let us rewrite the problem (A.1), (A.2) in the polar coordinates:

$$u''(r) + \frac{1}{r} u'(r) - q(r)u = 0, \qquad (\text{A.3})$$

$$u'(0) = 0, \quad u(1) = 1. \qquad (\text{A.4})$$

We consider it separately for $r < \delta$ and $r > \delta$. Let us introduce functions $v(r) = u(r)$ for $r < \delta$ and $w(r) = u(r)$ for $r > \delta$. Then, we have two problems:

$$v''(r) + \frac{1}{r} v'(r) - \omega v = 0, \quad 0 < r < \delta, \tag{A.5}$$

$$v'(0) = 0, \tag{A.6}$$

and

$$w''(r) + \frac{1}{r} w'(r) = 0, \quad \delta < r < 1, \tag{A.7}$$

$$w(1) = 1, \tag{A.8}$$

with the matching conditions

$$v(\delta) = w(\delta), \quad v'(\delta) = w'(\delta). \tag{A.9}$$

We use the notation $a = v(\delta)$.

Let us begin with the problem (A.7), (A.8). We have

$$(rw')' = 0.$$

Hence,

$$rw' = c_1, \quad w(r) = c_1 \ln r + c_2.$$

From (A.8), we obtain $c_2 = 1$, and from (A.9),

$$c_1 \ln \delta + 1 = a, \quad c_1 = \frac{a - 1}{\ln \delta}.$$

Hence,

$$w(r) = \frac{a - 1}{\ln \delta} \ln r + 1.$$

Let us consider the problem (A.5), (A.6). Introduce a new independent variable $\xi = \sqrt{\omega} r$ and a new function $z(\xi) = v(r)$. Then,

$$v'(r) = \sqrt{\omega}\, z'(\xi), \quad v''(r) = \omega\, z''(\xi),$$

$$z''(\xi) + \frac{1}{\xi} z'(\xi) - z(\xi) = 0, \quad 0 < \xi < \delta\sqrt{\omega}, \tag{A.10}$$

$$z'(0) = 0. \tag{A.11}$$

Let $\zeta = i\xi$ $(i = \sqrt{-1})$. Then,

$$y''(\zeta) + \frac{1}{\zeta} y'(\zeta) + y(\zeta) = 0, \quad 0 < |\zeta| < \delta\sqrt{\omega}. \qquad (A.12)$$

We have

$$y(\zeta) = c_3 J_0(\zeta),$$

where J_0 is the Bessel function of order zero. We use its known asymptotic expansion:

$$J_0(\zeta) = 1 - \frac{\zeta^2}{2^2} + \frac{\zeta^4}{2^2 \cdot 4^2} - \cdots . \qquad (A.13)$$

Solution to the problem (A.10), (A.11) reads

$$z(\xi) = y(\zeta) = c_3 J_0(i\xi) = c_3 \left(1 + \frac{\xi^2}{2^2} + \frac{\xi^4}{2^2 \cdot 4^2} + \cdots \right).$$

We can find c_3 and a from conditions (A.9):

$$c_3 J_0(i\sqrt{\omega}\delta) = a, \quad c_3\sqrt{\omega}iJ_0'(i\sqrt{\omega}\delta) = \frac{a-1}{\delta \ln \delta},$$

i.e.

$$c_3 J_0(i/\sqrt{\pi}) = a, \quad c_3 i J_0'(i/\sqrt{\pi}) = \frac{(a-1)\sqrt{\pi}}{\ln \delta},$$

or

$$a = \frac{\gamma}{\gamma - 1}, \quad (0 < a < 1),$$

where $\gamma = \dfrac{J_0(i/\sqrt{\pi})\sqrt{\pi}}{iJ_0'(i/\sqrt{\pi}) \ln \delta}$.

So, in conclusion,

$$u = \frac{a-1}{\ln \delta} \ln r + 1, \delta < r < 1, \quad u = \frac{a}{J_0(i\sqrt{\pi})} J_0(i\sqrt{\omega}r), 0 < r < \delta.$$
$$(A.14)$$

Note that a tends to zero as $\delta \to 0$. So, the solution tends to $u = 1$ almost everywhere. So, the almost-everywhere limit of the solution to (A.14) is not a solution to the equation

$$\Delta u - \delta(\mathbf{x}, \mathbf{0})u = 0,$$

where $\mathbf{x} = (x_1, x_2)$. However, the sequence u_δ, $\delta \to 0^+$, defined by (A.14), can be interpreted as a solution in the unitary disc of this equation (with the boundary condition (A.2)).

2. Periodic Dirac-like Potential

Consider a model of a periodic set of small cells in a nutrient. It is presented by the reaction–diffusion problem describing diffusion and consumption of nutrients in a biological tissue consisting of small cells periodically arranged in an extracellular matrix. Cells consume nutrients at a rate proportional to cell area and nutrient concentration. The dependence on the nutrient concentration can be linear or nonlinear. The cells are modeled by a potential approximating the Dirac's delta function. It depends on a large parameter w. Namely, it is equal to w on a periodic set of small balls having total measure of order w^{-1} (per unit of volume), and it is equal to zero out of the balls. We discuss the possibility of an asymptotic passage to the homogenized model in which the distance between the cells is a small parameter ε. The parameter ε may interact with w, and consequently for some values of the parameters, the problem becomes non-homogenizable in the classical sense. For the readers' convenience, we start with the case where the dependence on the nutrient concentration is linear and then generalize the construction in the case where it is nonlinear. This section contains the results [1].

Consider a small positive parameter ε, inverse to some integer number. First, we consider the linear problem

$$\frac{\partial u_{\varepsilon,w}}{\partial t} - \Delta u_{\varepsilon,w} + w q_\delta \left(\frac{x}{\varepsilon}\right) u_{\varepsilon,w} = f(x,t), \quad x \in \mathbb{R}^d, t > 0, \quad \text{(A.15)}$$

$$u_{\varepsilon,w}(x,0) = 0, \quad \text{(A.16)}$$

where $d = 2, 3$, and

$$q_\delta(\xi) = \begin{cases} 1, & \text{if } \xi \in B_\delta + \mathbb{Z}^d, \\ 0, & \text{otherwise.} \end{cases} \quad \text{(A.17)}$$

Here, B_δ is the ball $\{x \in \mathbb{R}^d, |x| < \delta\}$ such that $\text{mes} B_\delta = \dfrac{1}{w}$. So,

$$\text{mes} B_1 \delta^d = \frac{1}{w}. \quad \text{(A.18)}$$

For example, if $d = 2$, $\text{mes} B_\delta = \pi \delta^2 = \dfrac{1}{w}$ and $\delta = \sqrt{\dfrac{1}{w\pi}}$.

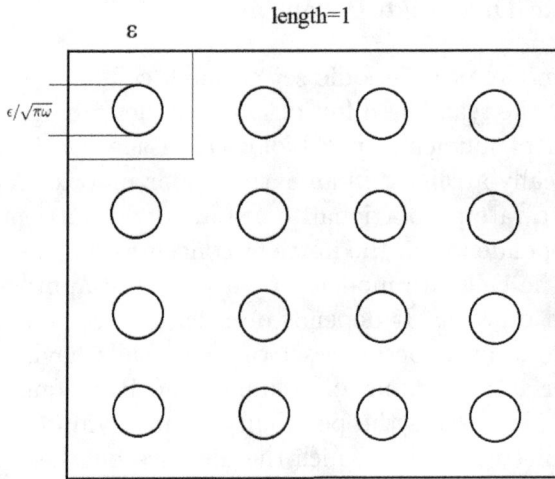

Figure 1.　Schematic representation of cells in unit-square domain.

We suppose that f is 1-periodic in x function, $f \in C^1(\mathbb{R}^d \times [0,T])$. The potential $\omega q_\delta \left(\dfrac{x}{\varepsilon} \right)$ models a δ-function with support in an ε-network $\varepsilon \mathbb{Z}^d$.

This model describes the distribution of the concentration of some biochemical substance (hormones, nutrients, drugs) in the presence of a periodic cell network, consuming the substance (Figure 1). Cells consume this substance at a rate proportional to its concentration and to the cell area or volume. The problem depends on two small parameters: ε and $\dfrac{1}{\omega}$ (measure of the potential support). We prove that if $\varepsilon^2 \omega \to 0$ and $\omega(\varepsilon^2 \omega)^J \to 0$ for some $J \in \mathbb{N}\backslash\{0\}$ when $\varepsilon \to 0, \omega \to +\infty$, then $u_{\varepsilon,\omega}$ tends to the solution for the limit problem

$$\frac{\partial v}{\partial t} - \Delta v + v = f(x,t), \quad x \in \mathbb{R}^d, t > 0,$$

$$v(x,0) = 0.$$

Then, we consider the nonlinear version of this problem:

$$\frac{\partial u_{\varepsilon,\omega}}{\partial t} - \Delta u_{\varepsilon,\omega} + \omega q_\delta \left(\frac{x}{\varepsilon} \right) F\left(u_{\varepsilon,\omega} \right) = f(x,t), \quad x \in \mathbb{R}^d, t > 0,$$

$$u_{\varepsilon,\omega}(x,0) = 0,$$

where F is $J + 2$ times continuously differentiable function with bounded derivatives of orders $j = 0, 1, \ldots, J + 2$, $F(0) = 0$, and $F'(u) > 0$ for all $u \in \mathbb{R}$, and as for the linear case, we prove that under the same conditions on ε and ω, $u_{\varepsilon,\omega}$ tends to the solution for the limit problem

$$\frac{\partial v}{\partial t} - \Delta v + F(v) = f(x, t), \quad x \in \mathbb{R}^d, t > 0,$$

$$v(x, 0) = 0.$$

Finally, we compare the theoretical results with numerical simulations carried out by L. Ait Mahiout.

It will be proved that the potential can be homogenized in the case where the condition $\varepsilon^2 \omega \ll 1$ is satisfied. If this condition is not satisfied, we cannot guarantee the closeness of the homogenized and initial models. In the case of the third dimension, we prove that if $\varepsilon \omega^{1/6} \gg 1$, then the solution for the initial model does not converge to the solution for the homogenized one. Moreover, outside of some small balls having total measure $1/\omega \to 0$, the solution for the original problem is close to the solution for the problem without potential (just the Laplace equation). We study numerically what happens in the case of violation of the condition $\varepsilon^2 \omega \ll 1$.

Note that this problem is close in some sense to the problem of homogenization of a perforated medium with periodic set of small holes with the Dirichlet's type boundary condition on the boundary of the holes solved by different methods in [4] and [3]. The Dirichlet's boundary condition on the holes can be interpreted as the value of $\omega = \infty$ in (A.15), (A.16). However, this case doesn't correspond to the Dirac's potential, and the result is quite different, because then there is only a weak convergence of the exact solution to the solution of the homogenized problem. Recall that below we call a problem "homogenizable" only in the case of strong convergence in some natural norm.

Assume that $\omega(\varepsilon^2 \omega)^J \to 0$ for some $J \in \mathbb{N} \backslash \{0\}$ when $\varepsilon \to 0, \omega \to +\infty$. Note that this assumption implies that $\varepsilon^2 \omega \to 0$. Consider the problem (A.15), (A.16).

Let T be a positive number independent of small parameters. Let f be 1-periodic in x function from $C^1(\mathbb{R}^d \times [0, T])$. Consider the problem (A.15), (A.16) for $t \leq T$. The variational formulation of this problem is well known. Let $H^1_{\#}$ ($L^2_{\#}$) be the space of

1-periodic functions $\mathbb{R}^d \rightarrow \mathbb{R}$ belonging to H^1_{loc} (L^2_{loc}), i.e. belonging to $H^1(B_R)$ ($L^2(B_R)$) for all $R > 0$. A weak solution for the problem (A.15), (A.16) is a function $u_{\varepsilon,\omega} \in L^\infty(0,T;H^1_\#)$ such that $\frac{\partial u_{\varepsilon,\omega}}{\partial t} \in L^2(0,T;L^2_\#)$, $u_{\varepsilon,\omega}|_{t=0} = 0$ and $\forall \psi \in H^1_\#$, for almost all $t \in (0,T)$,

$$\int_{(0,1)^d} \frac{\partial u_{\varepsilon,\omega}}{\partial t}(x,t)\psi(x) + \nabla u_{\varepsilon,\omega} \cdot \nabla \psi(x) + \omega q_\delta\left(\frac{x}{\varepsilon}\right) u_{\varepsilon,\omega}\psi(x)dx$$

$$= \int_{(0,1)^d} f(x,t)\psi(x)dx.$$

Here we use the Banach spaces $L^p(0,T;H)$. These spaces are the spaces of functions of time defined on $(0,T)$ and having values in a Hilbert space H.

For the considered coefficients and a smooth right-hand side f, this solution exists and is unique and satisfies the equation (A.15) pointwise everywhere except for the surfaces $\partial(B_{\delta\varepsilon} + \varepsilon\mathbb{Z}^d) \times (0,T)$, where the jump through the surface of $u_{\varepsilon,\omega}$ and $\frac{\partial u_{\varepsilon,\omega}}{\partial n}$ is equal to zero.

Let us introduce the limit problem

$$\frac{\partial v}{\partial t} - \Delta v + v = f(x,t), \quad x \in \mathbb{R}^d, t > 0, \tag{A.19}$$

$$v(x,0) = 0. \tag{A.20}$$

The following error estimate holds.

Let V be the Banach space of functions of $L^2(0,T;L^2_\#)$ having finite norm

$$\|u\|^2_V = \operatorname*{ess\,sup}_{t\in[0,T]} \|u(\cdot,t)\|^2_{L^2((0,1)^d)} + \|\nabla u\|^2_{L^2((0,1)^d \times (0,T))}.$$

Theorem A.1. *For any* $J \in \mathbb{N}\backslash\{0\}$,

$$\|u_{\varepsilon,\omega} - v\|_V = \mathcal{O}(\varepsilon\sqrt{\omega} + \sqrt{\omega}(\varepsilon^2\omega)^J). \tag{A.21}$$

Proof. In what follows, we use the standard calculus for matched expansions depending on the "slow" variable x and "fast" variable $\xi = x/\varepsilon$. Namely, we use the chain rule for products of differentiable functions N depending on the fast variable and V depending on the slow one:

$$\frac{\partial}{\partial x_i}(N(x/\varepsilon)V(x)) = \{\varepsilon^{-1}\frac{\partial N}{\partial \xi_i}(\xi)V(x) + N(\xi)\frac{\partial V}{\partial x_i}(x)\}|_{\xi=x/\varepsilon}.$$

Note that in this rule, the variables ξ and x are considered as *independent* between the brackets $\{$ and $\}$, while after the calculation of the whole expression, the restriction $\xi = x/\varepsilon$ is applied. In this sense, we omit the arguments of some functions.

Consider the following approximation:

$$u_{\varepsilon,\omega}^{(J)} = \left(1 + \varepsilon^2\omega N_{\delta,1}\left(\frac{x}{\varepsilon}\right) + \cdots + (\varepsilon^2\omega)^J N_{\delta,J}\left(\frac{x}{\varepsilon}\right)\right)v(x,t), \quad (A.22)$$

where $N_{\delta,j}$ satisfy the equations

$$\Delta N_{\delta,j}(\xi) = q_\delta(\xi)N_{\delta,j-1}(\xi) - \langle q_\delta(\xi)N_{\delta,j-1}\rangle, \quad (A.23)$$

with 1-periodicity condition, $N_{\delta,0} = 1$, $\langle N_{\delta,j}\rangle = \delta_{j,0}$, $j > 0$. Here,

$$\langle \cdot \rangle = \int_{(0,1)^d} \cdot \, d\xi.$$

Let us prove that

$$\|N_{\delta,j}\|_{H^1((0,1)^d)} \le C\delta^{d/2} \ (j > 0), \quad (A.24)$$

where C is a constant independent of small parameters. For $j = 1$, we get $N_{\delta,1}$ to be a 1-periodic solution to the equation

$$\Delta_\xi N_{\delta,1} = q_\delta(\xi) - \langle q_\delta(\xi)\rangle, \quad (A.25)$$

where the right-hand side has an $L^2((0,1)^d)$ norm which is less than

$$\|q_\delta\|_{L^2((0,1)^d)} + \|\langle q_\delta\rangle\|_{L^2((0,1)^d)} = \mathcal{O}(\delta^{d/2} + \delta^d) = \mathcal{O}(\delta^{d/2}).$$

Consider the equation $\Delta_\xi N = \mathcal{F}(\xi)$ with $\mathcal{F}$ 1-periodic L^2_{loc} function such that $\langle \mathcal{F}\rangle = 0$. Then, N satisfies the following estimate:

$$\int_{(0,1)^d} \nabla_\xi N \cdot \nabla_\xi N d\xi = \int_{(0,1)^d} (-\mathcal{F}N)d\xi \le \|\mathcal{F}\|_{L^2((0,1)^d)}\|N\|_{L^2((0,1)^d)}.$$

If N is such that $\langle N \rangle = 0$, then applying the Poincaré inequality, we get

$$||N||^2_{L^2((0,1)^d)} \leq \frac{d}{2}||\nabla_\xi N||^2_{L^2((0,1)^d)},$$

so

$$||N||_{H^1((0,1)^d)} \leq \sqrt{\frac{d}{2}+1}\sqrt{\frac{d}{2}}||F||_{L^2((0,1)^d)}. \qquad (A.26)$$

Then, we get estimate (A.24) for $j = 1$.

If (A.24) holds for some $j-1, j > 1$, then the right-hand side of equation (A.23) has an $L^2((0,1)^d)$ norm which is less than

$$||q_\delta N_{\delta,j-1}||_{L^2((0,1)^d)} + ||\langle q_\delta N_{\delta,j-1} \rangle||_{L^2((0,1)^d)}$$

$$\leq ||N_{\delta,j-1}||_{L^2((0,1)^d)} + \sqrt{\int_{(0,1)^d}\left|\int_{(0,1)^d} q_\delta N_{\delta,j-1}\mathrm{d}\xi\right|^2 \mathrm{d}x}$$

$$\leq 2||N_{\delta,j-1}||_{L^2((0,1)^d)} = \mathcal{O}(\delta^{d/2}).$$

So, we get (A.24) for j.

Plug (A.22) in equation (A.15):

$$\frac{\partial}{\partial t}u^{(J)}_{\varepsilon,\omega} - \Delta u^{(J)}_{\varepsilon,\omega} + \omega q_\delta\left(\frac{x}{\varepsilon}\right)u^{(J)}_{\varepsilon,\omega}$$

$$= \frac{\partial v}{\partial t} - \Delta v + v + \left\langle \omega q_\delta \sum_{j=1}^{J}(\varepsilon^2\omega)^j N_{\delta,j} \right\rangle v$$

$$+ \left\{ \omega q_\delta \sum_{j=0}^{J}(\varepsilon^2\omega)^j N_{\delta,j}v - \left\langle \omega q_\delta \sum_{j=0}^{J}(\varepsilon^2\omega)^j N_{\delta,j} \right\rangle v \right.$$

$$\left. + \omega\left(-\Delta_\xi \sum_{j=1}^{J}(\varepsilon^2\omega)^{j-1}N_{\delta,j}\right)v \right\} - 2\omega\varepsilon\sum_{i=1}^{d}\sum_{j=1}^{J}(\varepsilon^2\omega)^{j-1}\frac{\partial N_{\delta,j}}{\partial\xi_i}\frac{\partial v}{\partial x_i}$$

$$- \left(\sum_{j=1}^{J}(\varepsilon^2\omega)^j N_{\delta,j}\right)\Delta v + \left(\sum_{j=1}^{J}(\varepsilon^2\omega)^j N_{\delta,j}\right)\frac{\partial v}{\partial t}$$

$$= \frac{\partial v}{\partial t} - \Delta v + v + R_{0,\varepsilon,\omega} + R_{1,\varepsilon,\omega} + R_{2,\varepsilon,\omega} + R_{3,\varepsilon,\omega},$$

where

$$R_{0,\varepsilon,\omega} = \left\langle \omega q_\delta \sum_{j=1}^{J} (\varepsilon^2\omega)^j N_{\delta,j} \right\rangle v,$$

$$R_{1,\varepsilon,\omega} = \omega q_\delta \sum_{j=0}^{J} (\varepsilon^2\omega)^j N_{\delta,j} v - \left\langle \omega q_\delta \sum_{j=0}^{J} (\varepsilon^2\omega)^j N_{\delta,j} \right\rangle v$$

$$+ \omega \left(-\Delta_\xi \sum_{j=1}^{J} (\varepsilon^2\omega)^{j-1} N_{\delta,j} \right) v$$

$$= \omega \sum_{j=0}^{J-1} (\varepsilon^2\omega)^j \left\{ -\Delta_\xi N_{\delta,j+1} + q_\delta N_{\delta,j} - \langle q_\delta N_{\delta,j} \rangle \right\} v$$

$$+ \omega (\varepsilon^2\omega)^J (q_\delta N_{\delta,J} - \langle q_\delta N_{\delta,J} \rangle) v,$$

$$R_{2,\varepsilon,\omega} = -2\omega\varepsilon \left(\sum_{i=1}^{d} \sum_{j=1}^{J} (\varepsilon^2\omega)^{j-1} \frac{\partial N_{\delta,j}}{\partial \xi_j} \right) \frac{\partial v}{\partial x_i},$$

$$R_{3,\varepsilon,\omega} = \left(\sum_{j=1}^{J} (\varepsilon^2\omega)^j N_{\delta,j} \right) \left(\frac{\partial v}{\partial t} - \Delta v \right)$$

$$= \left(\sum_{j=1}^{J} (\varepsilon^2\omega)^j N_{\delta,j} \right) (-v + f).$$

Equation (A.19) is independent of small parameters, f is smooth, so v and $\dfrac{\partial v}{\partial x_i}$ are bounded by a constant independent of small parameters. $N_{\delta,j+1}$ satisfies equation (A.23), so all terms between {·} parentheses in the last equality satisfied by $R_{1,\varepsilon,\omega}$ vanish, and so

$$R_{1,\varepsilon,\omega} = \omega (\varepsilon^2\omega)^J (q_\delta N_{\delta,J} - \langle q_\delta N_{\delta,J} \rangle) v.$$

As before, we can prove that

$$\|q_\delta N_{\delta,J} - \langle q_\delta N_{\delta,J} \rangle\|_{L^2((0,1)^d)} = \mathcal{O}(\delta^{d/2}),$$

and so

$$\|R_{1,\varepsilon,\omega}\|_{L^2((0,1)^d\times(0,T))} = \mathcal{O}(\omega(\varepsilon^2\omega)^J \delta^{d/2}) = \mathcal{O}(\sqrt{\omega}(\varepsilon^2\omega)^J).$$

Taking into account that

$$|\langle q_\delta N_{\delta,j}\rangle| \leq \|q_\delta\|_{L^2((0,1)^d)}\|N_{\delta,j}\|_{L^2((0,1)^d)}$$

$$\leq \mathcal{O}(\delta^{d/2})\mathcal{O}(\delta^{d/2}) = \mathcal{O}(\delta^d) = \mathcal{O}\left(\frac{1}{\omega}\right),$$

we get

$$\|R_{0,\varepsilon,\omega}\|_{L^2((0,1)^d\times(0,T))} = \mathcal{O}(\varepsilon^2\omega).$$

Next, from the equalities

$$\left\|\frac{\partial N_{\delta,j}}{\partial\xi_i}\right\|_{L^2((0,1)^d)} = \mathcal{O}(\delta^{d/2}) = \mathcal{O}\left(\frac{1}{\sqrt{\omega}}\right),$$

we see that

$$\|R_{2,\varepsilon,\omega}\|_{L^2((0,1)^d\times(0,T))} = \mathcal{O}(\varepsilon\omega)\mathcal{O}\left(\frac{1}{\sqrt{\omega}}\right) = \mathcal{O}(\varepsilon\sqrt{\omega}).$$

Finally, $\|R_{3,\varepsilon,\omega}\|_{L^2((0,1)^d\times(0,T))} = \mathcal{O}(\varepsilon^2\omega)$.

Consider the difference $w = u_{\varepsilon,\omega} - u_{\varepsilon,\omega}^{(J)}$. It is a solution for the problem

$$\frac{\partial w}{\partial t} - \Delta w + \omega q_\delta\left(\frac{x}{\varepsilon}\right)w = R_{\varepsilon,\omega}(x,t), \quad x \in \mathbb{R}^d, t > 0, \qquad (A.27)$$

$$w(x,0) = 0, \qquad (A.28)$$

where $R_{\varepsilon,\omega} = -\sum_{i=0}^3 R_{i,\varepsilon,\omega}$,

$$\|R_{\varepsilon,\omega}\|_{L^2((0,1)^d\times(0,T))} = \mathcal{O}(\varepsilon\sqrt{\omega} + \sqrt{\omega}(\varepsilon^2\omega)^J). \qquad (A.29)$$

Multiplying (A.27) by w and integrating the product in $x \in (0,1)^d$ and $t \in [0,t]$, we get for any $t \in (0,T)$,

$$\frac{1}{2}\int_{(0,1)^d} w^2(x,t)\mathrm{d}x + \int_0^t\int_{(0,1)^d}\left(\nabla w \cdot \nabla w + \omega q_\delta\left(\frac{x}{\varepsilon}\right)w^2\right)\mathrm{d}x\mathrm{d}t$$

$$= \int_0^t\int_{(0,1)^d} R_{\varepsilon,\omega}w\ \mathrm{d}x\mathrm{d}t$$

$$\leq \|R_{\varepsilon,\omega}\|_{L^2((0,1)^d\times(0,T))}\|w\|_{L^2((0,1)^d\times(0,T))}$$

$$\leq \sqrt{2T}\|R_{\varepsilon,\omega}\|_{L^2((0,1)^d\times(0,T))}\|w\|_V. \qquad (A.30)$$

We used here the inequality $||w||_{L^2((0,1)^d \times (0,T))} \leq \sup_{t \in [0,T]} ||w(\cdot, t)||_{L^2((0,1)^d)} \sqrt{T}$. Then, passing to supremum for all $t \in [0,T]$, we get

$$||w||_V^2 \leq 2\sqrt{2T} ||R_{\varepsilon,w}||_{L^2((0,1)^d \times (0,T))} ||w||_V,$$

and so

$$||w||_V = \mathcal{O}(\varepsilon\sqrt{\omega} + (\varepsilon^2 \omega)^J \sqrt{\omega}).$$

Note that

$$||\nabla_\xi N_{j,\delta}||_{L^2((0,1)^d \times (0,T))} = \mathcal{O}(\delta^{d/2}) = \mathcal{O}\left(\frac{1}{\sqrt{\omega}}\right),$$

so

$$||u_{\varepsilon,\omega} - v||_V = \mathcal{O}(\varepsilon\sqrt{\omega} + (\varepsilon^2 \omega)^J \sqrt{\omega}).$$

The theorem is proved.

In the same way, one can consider a nonlinear problem:

$$\frac{\partial u_{\varepsilon,\omega}}{\partial t} - \Delta u_{\varepsilon,\omega} + \omega q_\delta \left(\frac{x}{\varepsilon}\right) F(u_{\varepsilon,\omega}) = f(x,t), \quad x \in \mathbb{R}^d, t > 0, \quad (\Lambda.31)$$

$$u_{\varepsilon,\omega}(x,0) = 0, \quad (A.32)$$

for a function F which is a *(H1) J+2 times continuously differentiable function with bounded derivatives of orders $j = 0, 1, \ldots, J+2$, $F(0) = 0$, and $F'(u) > 0$ for all $u \in \mathbb{R}$.*

As before, f is 1-periodic in x function from $C^1(\mathbb{R}^d \times [0,T])$.
The limit problem has the form

$$\frac{\partial v}{\partial t} - \Delta v + F(v) = f(x,t), \quad x \in \mathbb{R}^d, t > 0, \quad (A.33)$$

$$v(x,0) = 0. \quad (A.34)$$

For solutions to problems (A.31)–(A.34), estimate (A.21) is proved in Ref. [1].

2.1. Counterexample of the convergence of the solution for problem (A.15), (A.16) to the solution for the homogenized equation (A.19), (A.20) for $d = 3, \varepsilon\omega^{1/6} \gg 1$

We prove that if $d = 3, \varepsilon\omega^{1/6} \gg 1$, then there is no convergence of the solution for problem (A.15), (A.16) to the solution for the homogenized problem (A.19), (A.20). To this end, we prove that the solution for problem (A.15), (A.16) in this case converges to the solution for the equation without potential out of small neighborhoods of the discs, and this function is not close to the solution for the homogenized equation. In this section, f is supposed to be smoother: $f \in C^2(\mathbb{R}^3 \times [0, T])$ and additionally we assume that $f(x, 0) = 0$.

Consider the function $\eta \in C^2(\mathbb{R})$ defined by

$$\eta(x) = \begin{cases} 1 & \text{if } x \geq 2, \\ 0 & \text{if } x \leq 1. \end{cases} \tag{A.35}$$

Define now a 1-periodic function $\tilde{\eta} \in C^2(\mathbb{R}^3)$ such that

$$\tilde{\eta}_\omega(\xi) = \eta\left(\frac{|\xi|}{\delta}\right), \ \forall \xi \in \left(-\frac{1}{2}, \frac{1}{2}\right)^3, \tag{A.36}$$

where $\delta = \left(\frac{4\omega\pi}{3}\right)^{-1/3}$ is the radius of a ball B_δ having measure $\frac{1}{\omega}$, $\delta < \frac{1}{2}$.

Consider

$$u^a(x, t) = \tilde{w}(x, t)\tilde{\eta}_\omega\left(\frac{x}{\varepsilon}\right) + \left(1 - \tilde{\eta}_\omega\left(\frac{x}{\varepsilon}\right)\right)\frac{f(x, t)}{\omega}, \tag{A.37}$$

where $\tilde{w}$ is a solution for the problem

$$\begin{cases} \dfrac{\partial \tilde{w}}{\partial t} - \Delta\tilde{w} = f(x, t), & x \in \mathbb{R}^3, t > 0, \\ \tilde{w}(x, t = 0) = 0. \end{cases} \tag{A.38}$$

Then, for $x \in B_{\varepsilon\delta} + (\varepsilon\mathbb{Z})^3$, we get

$$\frac{\partial u^a}{\partial t} - \Delta u^a + \omega q_\delta\left(\frac{x}{\varepsilon}\right)u^a = f(x, t) + \frac{1}{\omega}\left(\frac{\partial f}{\partial t} - \Delta f\right) = f + R_{\varepsilon,\omega}(x, t), \tag{A.39}$$

with $\|R_{\varepsilon,\omega}\|_{L^2((0,1)^3 \times (0,T))} = \mathcal{O}\left(\frac{1}{\omega}\right)$.

For $x \in \mathbb{R}^3 \setminus (B_{2\varepsilon\delta} + (\varepsilon\mathbb{Z}^3))$, we get

$$\frac{\partial u^a}{\partial t} - \Delta u^a + \omega q_\delta \left(\frac{x}{\varepsilon}\right) u^a = \frac{\partial u^a}{\partial t} - \Delta u^a = \frac{\partial \tilde{w}}{\partial t} - \Delta \tilde{w} = f, \quad (\text{A.40})$$

and for $x \in (B_{2\varepsilon\delta} \setminus B_{\varepsilon\delta}) + (\varepsilon\mathbb{Z})^3$, we get

$$\frac{\partial u^a}{\partial t} - \Delta u^a + \omega q_\delta \left(\frac{x}{\varepsilon}\right) u^a = \frac{\partial u^a}{\partial t} - \Delta u^a = f + R_{\varepsilon,\omega}(x,t), \quad (\text{A.41})$$

where

$$\begin{aligned}
R_{\varepsilon,\omega}(x) &= -f(x,t) + \frac{\partial \tilde{w}}{\partial t} \tilde{\eta}_\omega \left(\frac{x}{\varepsilon}\right) + \frac{1}{\omega}\frac{\partial f}{\partial t}\left(1 - \tilde{\eta}_\omega\left(\frac{x}{\varepsilon}\right)\right) \\
&\quad - \operatorname{div}\left\{\nabla\left(\tilde{w}\tilde{\eta}_\omega\left(\frac{x}{\varepsilon}\right)\right) + \nabla\left(\left(1 - \tilde{\eta}_\omega\left(\frac{x}{\varepsilon}\right)\right)f\right)\omega^{-1}\right\} \\
&= -f(x,t) + \left(\frac{\partial \tilde{w}}{\partial t} - \Delta\tilde{w}\right)\tilde{\eta}_\omega\left(\frac{x}{\varepsilon}\right) + \frac{1}{\omega}\frac{\partial f}{\partial t}\left(1 - \tilde{\eta}_\omega\left(\frac{x}{\varepsilon}\right)\right) \\
&\quad - (\nabla\tilde{w})\cdot\left(\nabla\tilde{\eta}_\omega\left(\frac{x}{\varepsilon}\right)\right) - \operatorname{div}\left(\tilde{w}\nabla\tilde{\eta}_\omega\left(\frac{x}{\varepsilon}\right)\right) \\
&\quad - \operatorname{div}\left(\nabla\left(\left(1 - \tilde{\eta}_\omega\left(\frac{x}{\varepsilon}\right)\right)f\right)\right)\omega^{-1}.
\end{aligned}$$

Here,

$$-f + \left(\frac{\partial \tilde{w}}{\partial t} - \Delta\tilde{w}\right)\tilde{\eta}_\omega\left(\frac{x}{\varepsilon}\right) = f\left(\tilde{\eta}_\omega\left(\frac{x}{\varepsilon}\right) - 1\right),$$

and its support in $(0,1)^3 \times (0,T)$ has a measure of order $\dfrac{1}{\omega}$, and measures of the supports of all other terms

$$\frac{1}{\omega}\frac{\partial f}{\partial t}\left(1 - \tilde{\eta}_\omega\left(\frac{x}{\varepsilon}\right)\right), \quad (\nabla\tilde{w})\cdot\left(\nabla\tilde{\eta}_\omega\left(\frac{x}{\varepsilon}\right)\right), \quad \operatorname{div}\left(\tilde{w}\nabla\tilde{\eta}_\omega\left(\frac{x}{\varepsilon}\right)\right),$$

$$\operatorname{div}\left(\nabla\left(\left(1 - \tilde{\eta}_\omega\left(\frac{x}{\varepsilon}\right)\right)f\right)\right)\omega^{-1}$$ are also of order $\dfrac{1}{\omega}$; on the other hand, $|\nabla\tilde{\eta}_\omega| = \mathcal{O}\left(\dfrac{\omega^{1/3}}{\varepsilon}\right)$, and so

$$\|\tilde{w}\nabla\tilde{\eta}_{\omega(\xi)}(\cdot/\varepsilon)\|_{L^2((0,1)^3\times(0,T))} = \sqrt{\frac{\omega^{2/3}}{\varepsilon^2}\frac{1}{\omega}} = \frac{1}{\varepsilon(\omega)^{1/6}}. \quad (\text{A.42})$$

So finally, for $x \in \mathbb{R}^3$,

$$\frac{\partial u^a}{\partial t} - \Delta u^a + \omega q_\delta \left(\frac{x}{\varepsilon}\right) u^a = f + R_{\varepsilon,\omega}(x,t), \qquad (A.43)$$

$$R_{\varepsilon,\omega} = R^0_{\varepsilon,\omega} - \operatorname{div} R^1_{\varepsilon,\omega}, \qquad (A.44)$$

where

$$R^0_{\varepsilon,\omega}(x,t) =$$

$$\begin{cases} \dfrac{1}{\omega}\left(\dfrac{\partial f}{\partial t} - \Delta f\right) & \text{in } B_{\varepsilon\delta} + (\varepsilon\mathbb{Z}^3), \\[2mm] f\left(\tilde{\eta}_\omega\left(\dfrac{x}{\varepsilon}\right) - 1\right) + \dfrac{1}{\omega}\dfrac{\partial f}{\partial t}\left(1 - \tilde{\eta}_\omega\left(\dfrac{x}{\varepsilon}\right)\right) \\[2mm] \quad -(\nabla\tilde{w})\cdot(\nabla\tilde{\eta}_\omega) & \text{in } \mathbb{R}^3 \setminus (B_{\varepsilon\delta} + (\varepsilon\mathbb{Z}^3)), \end{cases} \qquad (A.45)$$

$$R^1_{\varepsilon,\omega} = \tilde{w}\nabla\tilde{\eta}_\omega\left(\frac{x}{\varepsilon}\right) + \nabla((1 - \tilde{\eta}_\omega)f)\omega^{-1}, \qquad (A.46)$$

$$\|R^0_{\varepsilon,\omega}\|_{L^2((0,1)^3\times(0,T))} = \mathcal{O}\left(\frac{1}{\varepsilon\omega^{1/6}}\right), \qquad (A.47)$$

$$\|R^1_{\varepsilon,\omega}\|_{L^2((0,1)^3\times(0,T))} = \mathcal{O}\left(\frac{1}{\varepsilon\omega^{1/6}}\right), \qquad (A.48)$$

As before (see (A.27)–(A.30)), we can prove that $\|u^a - u_{\varepsilon,\delta}\|_V = \mathcal{O}\left(\dfrac{1}{\varepsilon\omega^{1/6}}\right)$; only one modification should be done: the right-hand side of (A.30) admits the following estimate:

$$\int_0^t \int_{(0,1)^3} (R^0_{\varepsilon,\omega}w + R^1_{\varepsilon,\omega}\cdot\nabla w)\,dxdt$$

$$\leq \|R^0_{\varepsilon,\omega}\|_{L^2((0,1)^3\times(0,T))}\|w\|_{L^2((0,1)^3\times(0,T))}$$

$$+ \|R^1_{\varepsilon,\omega}\|_{L^2((0,1)^3\times(0,T))}\|\nabla w\|_{L^2((0,1)^3\times(0,T))}$$

$$\leq \left(\sqrt{2T}\|R^0_{\varepsilon,\omega}\|_{L^2((0,1)^3\times(0,T))}\right.$$

$$\left. + \|R^1_{\varepsilon,\omega}\|_{L^2((0,1)^3\times(0,T))}\right)\|w\|_V.$$

So, if $\varepsilon\omega^{1/6} \to +\infty$, then in the norm V, $u_{\varepsilon,\delta}$ tends to $\tilde{w}$ (out of $B_{2\delta\varepsilon} + (\varepsilon\mathbb{Z})^3$) and not to v.

To confirm the theoretical results obtained above, we present numerical simulations in 2-D and 3-D cases using the finite element code FreeFem++ for three problems: heterogeneous problem, the corresponding homogenized problem and associated problem without potential. We compare solutions of these problems at time $t = 0.5$. We show that increasing of values of the parameter $\alpha = \varepsilon^2 \omega$ leads to the loss of "homogenizability" of the equation, and for large values of α, the solution starts to approach the solution to the problem without potential outside of small neighborhoods of the inclusions. This behavior is visible in the numerical tests; however, theoretically, we justify it only for sufficiently large values of α: if α is greater than $\omega^{2/3}$.

3. Numerical Tests in 2-D

Consider the following three problems:
  The nonlinear heterogeneous problem:

$$K \left(\frac{\partial u_{\varepsilon,\omega}}{\partial t} - \Delta u_{\varepsilon,\omega} \right) + \omega q_\delta \left(\frac{x}{\varepsilon}, \frac{y}{\varepsilon} \right) \frac{u_{\varepsilon,\omega}}{1 + u_{\varepsilon,\omega}} = Kt \cos(2\pi x); \tag{A.49}$$
$$(x, y) \in (0, 1)^2, \quad t \in [0, T],$$

where

$$q_\delta(t, s) = \begin{cases} 1, & \text{if } (t, s) \in B_\delta + \mathbb{Z}^2, \\ 0, & \text{otherwise.} \end{cases}$$

The diameter of each ball of support $q_\delta \left(\frac{x}{\varepsilon}, \frac{y}{\varepsilon} \right)$ is $2\varepsilon \dfrac{1}{\sqrt{\omega \pi}} \ll \varepsilon$.
  The initial condition is

$$u_{\varepsilon,\omega}(x, y, t = 0) = 0, \tag{A.50}$$

and for the space variables, we set the periodic boundary conditions.
  The homogenized problem:

$$K \left(\frac{\partial v}{\partial t} - \Delta v \right) + \frac{v}{1 + v} = Kt \cos(2\pi x), \quad (x, y) \in (0, 1)^2, t \in [0, T]; \tag{A.51}$$

$$v(x, y, 0) = 0, \tag{A.52}$$

with periodic boundary conditions. Although the condition of derivability of F is not valid for $v = -1$, it is satisfied for all values of $v > -1$, and the solution v of (A.51)–(A.52) is greater than -1.

The problem without potential:

$$\left(\frac{\partial w}{\partial t} - \Delta w\right) = Kt\cos(2\pi x); \quad (x,y) \in (0,1)^2, t \in [0,1]. \quad \text{(A.53)}$$

We consider the initial condition

$$w(x,y,t=0) = 0, \quad \text{(A.54)}$$

and periodic boundary conditions. For all three problems, we take $K = 0.025$.

Setting $\alpha = \varepsilon^2 \omega$, we compute a numerical solution for problems (A.49)–(A.50), (A.51)–(A.52), and (A.53)–(A.54) for different values of $\alpha : \alpha \ll 1$, $\alpha = 1$, and $\alpha \gg 1$. In each case, we present the graph of the approximate solutions u_α, v, and w as a function of x, for $y = 0$ fixed, at time $t = 0.5$.

Case 1: tests with values of ε and ω such that $\alpha \ll 1$.

- $\varepsilon = 0.0625, \omega = 24 : \alpha = 0.093$:
  Figure 2 shows the convergence of the solution u for the heterogeneous problem (A.49)–(A.50) to the solution v for the homogenized problem (A.51)–(A.52).

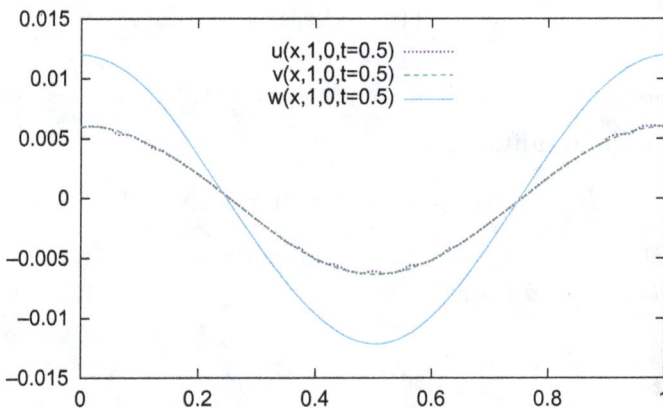

Figure 2. Graph of u_α, v, and w at time $t = 0.5$ of u_α as a function of x, at $y = 1$ fixed, for the case $\alpha = 0.093$.

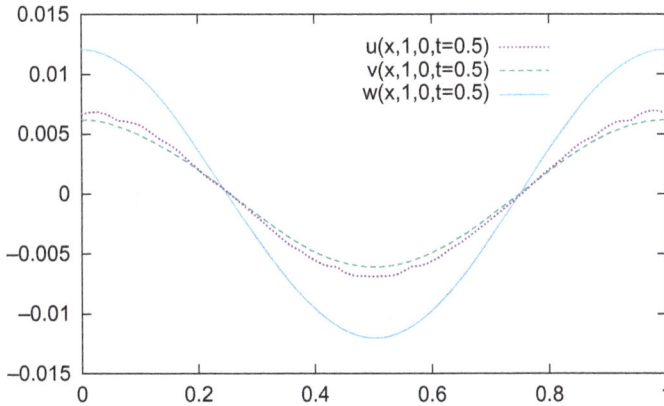

Figure 3. Graph of u_α, v, and w at time $t = 0.5$ as a function of x, at $y = 1$ fixed, for the case $\alpha = 1$.

Case 2: tests with values of ε and ω such that $\alpha = 1$.

• $\varepsilon = 0.0625, \omega = 256 : \alpha = 1$:

Figure 3 shows the appearance of oscillations of solution for the heterogeneous problem (A.49)–(A.50).

Case 3: tests with values of ε and ω such that $\alpha \gg 1$.

• $\varepsilon = 0.3, \omega = 10000 : \alpha = 900$:

Figure 4 illustrates the absence of convergences for the heterogeneous problem (A.49)–(A.50) to the homogenized problem (A.51)–(A.52) if $\alpha \gg 1$.

4. Numerical Tests in 3-D

Consider the following problems:
The heterogeneous problem:

$$K\left(\frac{\partial u_\varepsilon}{\partial t} - \Delta u_\varepsilon\right) + \omega q_\delta\left(\frac{x}{\varepsilon}, \frac{y}{\varepsilon}, \frac{z}{\varepsilon}\right) u_\varepsilon = (K(1 + 4t\pi^2) + t)\sin(2\pi x);$$

$$(x, y, z) \in (0, 1)^3, t \in [0, 1],$$

$$(A.55)$$

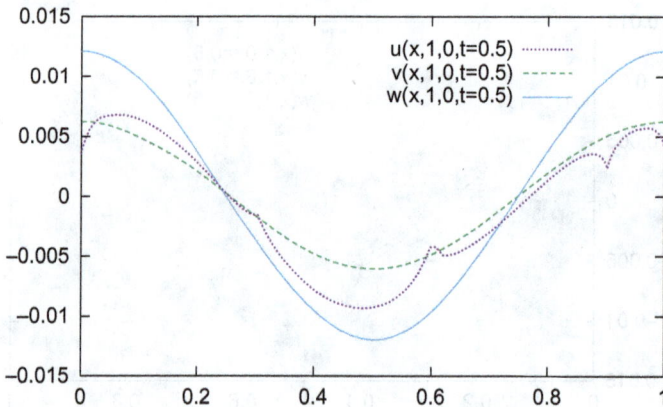

Figure 4. Graph of u_α, v, and w at time $t = 0.5$ as a function of x, at $y = 1$ fixed, for the case $\alpha = 6$.

where

$$q_\delta(s_1, s_2, s_3) = \begin{cases} 1, & \text{if } (s_1, s_2, s_3) \in B_\delta + \mathbb{Z}^3, \\ 0, & \text{otherwise.} \end{cases} \tag{A.56}$$

The diameter of each ball of support $q_\delta\left(\dfrac{x}{\varepsilon}, \dfrac{y}{\varepsilon}, \dfrac{z}{\varepsilon}\right)$ is $2\varepsilon\left(\dfrac{3}{4}\dfrac{1}{\omega\pi}\right)^{1/3} \ll \varepsilon.$

The initial condition is

$$u_\varepsilon(x, y, z, t = 0) = 0, \tag{A.57}$$

and the 1-periodic boundary conditions are set.

The homogenized problem:

$$K\left(\frac{\partial v}{\partial t} - \Delta v\right) + v = (K(1 + 4t\pi^2) + t)\sin(2\pi x);$$
$$(x, y, z) \in (0, 1)^3, t \in [0, 1]. \tag{A.58}$$

As before, the initial condition is

$$v(x, y, z, t = 0) = 0, \tag{A.59}$$

and the 1-periodic boundary conditions are set.

The problem without potential:

$$K \left(\frac{\partial w}{\partial t} - \Delta w \right) = (K(1 + 4t\pi^2) + t) \sin(2\pi x);$$

(A.60)

$$(x, y, z) \in (0,1)^3, t \in [0,1],$$

$$w(x, y, z, t = 0) = 0,$$

(A.61)

and periodic boundary conditions are considered.

Introducing notation $\beta = \varepsilon w^{1/6}$, we compute a numerical solution for problems (A.60)–(A.61), (A.55)–(A.57), and (A.58)–(A.59) for different values of β : $\beta \ll 1$, $\beta = 1$, and $\beta \gg 1$ noting that the condition $\beta \gg 1$ is equivalent to $\alpha \gg w^{2/3}$.

In each case, we present the graph of approximate solutions u_α, v, w as a function of x, for $y = 1$, $z = 0$ at time $t = 0.5$.

Case 1: tests with values of ε and w such that $\beta \ll 1$.

- $\varepsilon = 0.2, w = 2.31 : \beta = 0.23$:
  Figure 5 shows the convergence of the solution u for the heterogeneous problem (A.55)–(A.57) to the solution v for the homogenized problem (A.58)–(A.59).

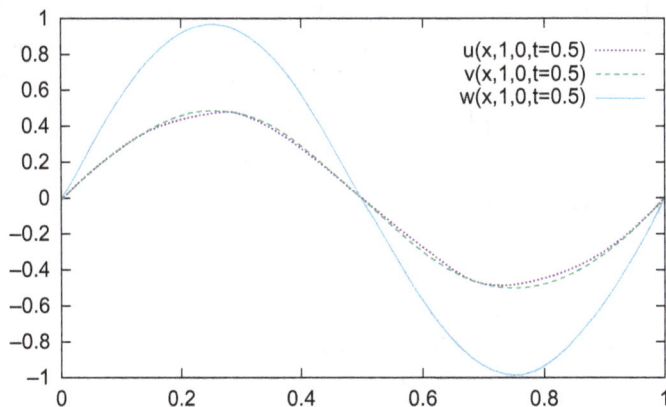

Figure 5. Graph of u_β, v, and w at time $t = 0.5$ as a function of x, at $y = 1$, $z = 0$ fixed, for the case $\beta = 0.23$.

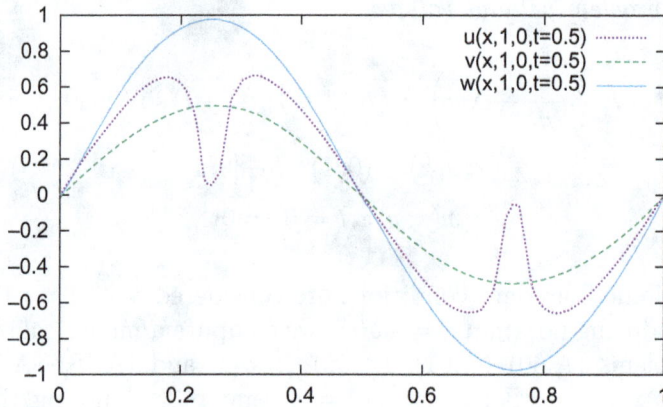

Figure 6.   Graph of u_β, v, and w at time $t = 0.5$ as a function of x, at $y = 1$, $z = 0$ fixed, for the case $\beta = 1$.

Case 2: tests with values of ε and ω such that $\beta = 1$.

- $\varepsilon = 0.25, \omega = 4096 : \beta = 1$:

Figure 6 shows the appearance of oscillations of solution for heterogeneous problem (A.49)–(A.50).

Case 3: tests with values of ε and ω such that $\beta \gg 1$.

- $\varepsilon = 0.3, \omega = 10000 : \beta = 2$:

Note that if $\beta = \varepsilon \omega^{1/6}$ is sufficiently large, the solution for problem (A.55)–(A.57) oscillates, and outside the balls, it approaches the solution for problem without potential (A.60)–(A.61) being far from the solution for problem (A.58)–(A.59) (see Figure 7).

5.   Partial Homogenization

The main idea of this section is that in the case where we have the potential satisfying the condition (H) $\varepsilon^2 \omega \ll 1$ only in a part of the domain, then we can partially homogenize the problem leaving the original potential in the part where this condition is not satisfied. It means that we can combine the homogenized and quasi-discrete description into one hybrid model. In the following we evaluate the difference between the solutions of the completely quasi-discrete and

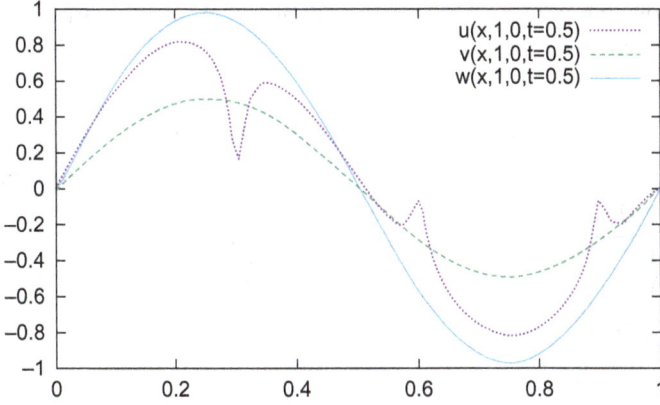

Figure 7. Graph of u_β, v, and w at time $t = 0.5$ as a function of x, at $y = 1$, $z = 0$ fixed, for the case $\beta = 2$.

partially homogenized problems is small. For the diffusion equation with rapidly oscillating coefficients the partial homogenization method was introduced first in Ref. [5].

Compare the following problems. The first one is the diffusion equation with the quasi-discrete potential:

$$
\begin{cases}
\dfrac{\partial u_{\varepsilon,\omega_1,\omega_2}}{\partial t} - \Delta u_{\varepsilon,\omega_1,\omega_2} + q_{\omega_1,\omega_2}\left(\dfrac{x}{\varepsilon}, x_1\right) u_{\varepsilon,\omega_1,\omega_2} \\
\qquad = f_{\varepsilon,\omega_1,\omega_2}(x,t), \quad x \in \mathbb{R}^d, t > 0, \\
u_{\varepsilon,\omega_1,\omega_2}(x,0) = 0,
\end{cases}
\tag{A.62}
$$

where $d = 2, 3$,

$$
q_{\omega_1,\omega_2}(\xi, x_1) =
\begin{cases}
\omega_1, & \text{if } \xi \in B_{\delta_1} + (\mathbb{Z} + 1/2)^d, x_1 \in [0, 1 - \mu) + \mathbb{Z}, \\
\omega_2, & \text{if } \xi \in B_{\delta_2} + (\mathbb{Z} + 1/2)^d, x_1 \in (1 - \mu, 1] + \mathbb{Z}, \\
0, & \text{otherwise.}
\end{cases}
\tag{A.63}
$$

B_{δ_i} is the ball $\{x \in \mathbb{R}^d, |x| < \delta_i\}$ such that $\mathrm{mes} B_{\delta_i} = \dfrac{1}{\omega_i}$. So,

$$
\mathrm{mes} B_1 \delta_i^d = \frac{1}{\omega_i}, \quad i = 1, 2,
\tag{A.64}
$$

μ is a positive constant independent of parameters ε and ω_i, $f_{\varepsilon,\omega_1,\omega_2}$ is 1-periodic in x function, $f_{\varepsilon,\omega_1,\omega_2} \in C^1(\mathbb{R}^d \times [0,T])$. This problem is called "original." We assume that

$$\varepsilon \to 0, \quad \omega_i \to +\infty, \quad i = 1,2, \quad \varepsilon^2\omega_1 \to 0.$$

The second problem is the partially homogenized one, i.e.

$$\begin{cases} \dfrac{\partial v_{\varepsilon,\omega_1,\omega_2}}{\partial t} - \Delta v_{\varepsilon,\omega_1,\omega_2} + \hat{q}_{\omega_1,\omega_2}\left(\dfrac{x}{\varepsilon}, x_1\right) v_{\varepsilon,\omega_1,\omega_2} \\ \qquad\qquad = f_{\varepsilon,\omega_1,\omega_2}(x,t), \quad x \in \mathbb{R}^d, t > 0, \qquad \text{(A.65)} \\ v_{\varepsilon,\omega_1,\omega_2}(x,0) = 0, \end{cases}$$

where

$$\hat{q}_{\omega_1,\omega_2}(\xi,x_1) = \begin{cases} 1, & \text{if } x_1 \in [0, 1-\mu) + \mathbb{Z}, \\ \omega_2, & \text{if } \xi \in B_{\delta_2} + \mathbb{Z}^d, x_1 \in (1-\mu, 1] + \mathbb{Z}, \quad \text{(A.66)} \\ 0, & \text{otherwise.} \end{cases}$$

So here, the potential $\hat{q}_{\omega_1,\omega_2}$ is homogenized in the part $x_1 \in [0, 1-\mu) + \mathbb{Z}$. Clearly, the method of partial homogenization is especially effective when μ is relatively small, i.e. when the area of the direct numerical computation, where the homogenization cannot be applied, is not excessively large. In what follows, assume that 1 and $1 - \mu$ are multiples of ε. The following theorem is proved in Ref. [2].

Theorem A.2. *Let M be the norm*

$$M = \|v_{\varepsilon,\omega_1,\omega_2}\|_{L^\infty((0,1)^d \times (0,T))} + \|\nabla_x v_{\varepsilon,\omega_1,\omega_2}\|_{L^\infty((0,1)^d \times (0,T))}$$
$$+ \|f_{\varepsilon,\omega_1,\omega_2}\|_{L^\infty((0,1)^d \times (0,T))}.$$

For any $J \in \mathbb{N}$,

$$\|u_{\varepsilon,\omega_1,\omega_2} - v_{\varepsilon,\omega_1,\omega_2}\|_{V((0,1)^d \times (0,T))} = \mathcal{O}(\sqrt{\varepsilon\omega_1} + \sqrt{\omega_1}(\varepsilon^2\omega_1)^J)M,$$
$$\text{(A.67)}$$

where

$$\|u\|_V^2 = \frac{1}{2} \sup_{t \in [0,T]} \|u(\cdot,t)\|_{L^2((0,1)^d)}^2 + \|\nabla u(\cdot,t)\|_{L^2((0,1)^d \times (0,T))}^2.$$

This theorem confirms that the error is small if the norm M of the solution for the semi-homogenized problem is bounded and if $\varepsilon\omega_1 \ll 1$. The last condition is more restrictive than the above-discussed condition (H) that $\varepsilon^2\omega_1 \ll 1$. However, the numerical experiments show that the method works for the less-restrictive condition (H).

Appendix B

Proof of Riesz–Frechet Representation Theorem

This appendix contains the proof [6] of the Riesz–Frechet representation theorem, see Theorem 1.6 at Section 1.4, Chapter 2.

Definition B.1. Let H be a Hilbert space, $x \in H$, $M \subset H$. Then, the distance between x and M is defined as

$$\text{dist}(x, M) = \inf_{u \in M} \|x - u\|.$$

Definition B.2. Let H be a vector space, $M \subset H$. M is convex if and only if $\forall (x, y) \in M^2$, $\forall t \in [0, 1]$, $tx + (1 - t)y \in M$.

Lemma B.1. *If $x \in M$, then* $\text{dist}(x, M) = 0$.

If $x \notin M$, and M is closed, then $\text{dist}(x, M) > 0$.

Proof. If $x \in M$, then taking $u = x$, we see that $\text{dist}(x, M) = 0$.

If $x \notin M$, we prove $\text{dist}(x, M) > 0$. Assume that $\text{dist}(x, M) = 0$, then $\forall \varepsilon_n > 0$, $\exists x_n \in M$ such that $\|x - x_n\| < \varepsilon_n$.

Let us choose $(\varepsilon_n)_{n \in \mathbb{N}}$ converging to zero.

Then, $x_n \to x$ $(n \to +\infty)$, and so $x \in M$ (M is closed). This implies $\text{dist}(x, M) > 0$.

The lemma is proved. $\square$

Theorem B.1. *Let H be a Hilbert space, and let M be a closed convex set.*

Then, $\forall x \notin M \; \exists! y \in M$ such that $\|x - y\| = \text{dist}(x, M)$.

Proof. Denote $d = \text{dist}(x, M) = \inf_{u \in M} \|x - u\|$. By the previous lemma, $d > 0$. So, by definition of inf, there exists a sequence $(u_n)_{n \in \mathbb{N}} \subset M$ such that

$$d \le \|x - u_n\| \le d + \frac{1}{n}. \tag{B.1}$$

Let us prove that the sequence $(u_n)_{n \in \mathbb{N}}$ is a Cauchy sequence. By the parallelogram identity, $\forall m, n \in \mathbb{N}$ for $x - u_n$ and $x - u_m$,

$$\|x - u_n + x - u_m\|^2 + \|x - u_n - x + u_m\|^2 = 2(\|x - u_n\|^2 + \|x - u_m\|^2),$$

$$2\|x - u_n\|^2 + 2\|x - u_m\|^2 = 4\left\|x - \frac{u_n + u_m}{2}\right\|^2 + \|u_n - u_m\|^2.$$

On the other hand, $\dfrac{u_n + u_m}{2} = \frac{1}{2} u_n + \frac{1}{2} u_m \in M$ as M is convex.

So, $\left\|x - \dfrac{u_n + u_m}{2}\right\|^2 \ge d^2$.

So, by (B.1),

$$\|x - u_n\|^2 \le \left(d + \frac{1}{n}\right)^2 \quad \text{and} \quad \|x - u_m\|^2 \le \left(d + \frac{1}{m}\right)^2.$$

So,

$$\|u_n - u_m\|^2 = 2\|x - u_n\|^2 + 2\|x - u_m\|^2 - 4\left\|x - \frac{u_n + u_m}{2}\right\|^2$$

$$\le 2\left(d + \frac{1}{n}\right)^2 + 2\left(d + \frac{1}{m}\right)^2 - 4d^2$$

$$= 4d\left(\frac{1}{n} + \frac{1}{m}\right) + 2\left(\frac{1}{n^2} + \frac{1}{m^2}\right).$$

So, $\forall \varepsilon > 0, \; \exists n_0 \in \mathbb{N}$ such that $n, m > n_0, \; \|u_n - u_m\|^2 < \varepsilon$.

So, $(u_n)_{n \in \mathbb{N}}$ is a Cauchy sequence. H is complete, so there exists $y \in H$ such that $u_n \to y$. M is a closed set, so $y \in M$.

Passing to the limit $m \to +\infty$ in (B.1), we get $x - u_n \to x - y$ and $\|x - y\| = d$.

So, we find $y \in M$ such that $\|x - y\| = \text{dist}(x, M)$.
Let us prove the uniqueness of y.
Let y^* be an element of M such that

$$\|x - y^*\| = \text{dist}(x, M) = d.$$

Applying the parallelogram identity, we get

$$4d^2 = 2\|x - y\|^2 + 2\|x - y^*\|^2 = 4 \left\| x - \frac{y + y^*}{2} \right\|^2 + \|y - y^*\|^2.$$

Note that $\dfrac{y + y^*}{2} \in M$ (M is convex), so

$$4 \left\| x - \frac{y + y^*}{2} \right\|^2 \geq 4d^2.$$

Consequently,

$$\|y - y^*\|^2 \leq 4d^2 - 4 \left\| x - \frac{y + y^*}{2} \right\|^2 \leq 4d^2 - 4d^2 = 0.$$

So, $\|y - y^*\|^2 \leq 0$ and $y = y^*$.
Therefore, y is unique. $\qquad \qquad \square$

Corollary B.1. *Let H be a Hilbert space, and let L be a Hilbert subspace. Let $x \notin L$.*

Then, $\exists! y \in L$ such that $\text{dist}(x, L) = \|x - y\|$.

Proof. It is enough to check that L is convex and closed. L is convex as a vector space.
Let us show that L is closed.
Let $(u_n)_{n \in \mathbb{N}}$ be a sequence converging to $u \in H$ such that $u_n \in L$. Then, $(u_n)_{n \in \mathbb{N}}$ is a Cauchy sequence the completeness L yields: $u_n \to w \in L$. Thus, $u = w \in L$. $\qquad \qquad \square$

Denote $L^{\perp} = \{x \in H; x \perp L\} = \{x \in H | \forall y \in L, (x, y) = 0\}$, where L is a subspace of H.

Theorem B.2. *Let H be an $\mathbb{R}$ — Hilbert space, L be a Hilbert subspace, and let y be an element of L such that $\text{dist}(x, L) = \|x - y\|$.*

Then, $x - y \perp L$, i.e. $\forall h \in L$, $((x - y), h) = 0$.

Proof. Let us prove that $\forall h \in L$, $((x - y), h) = 0$.
For any $\lambda \in \mathbb{R}$, we get

$$\|x - y + \lambda h\| \geq \|x - y\|.$$

Then,

$$\|x - y + \lambda h\|^2 \geq \|x - y\|^2,$$

i.e.

$$((x - y + \lambda h), (x - y + \lambda h)) \geq ((x - y), (x - y)),$$

i.e.

$$((x - y), (\lambda h)) + ((\lambda h), (x - y)) + |\lambda|^2 \|h\|^2 \geq 0,$$

i.e.

$$(\lambda(x - y), h) + (\lambda h, (x - y)) + |\lambda|^2 \|h\|^2 \geq 0.$$

Let us take

$$\lambda = -\frac{((x - y), h)}{\|h\|^2} \quad (h \neq 0).$$

Then,

$$-\frac{(h, (x - y))}{\|h\|^2}((x - y), h) - \frac{((x - y), h)}{\|h\|^2}h(x - y)$$

$$+ \frac{|((x - y), h)|^2}{\|h\|^4}\|h\|^2 \geq 0,$$

and so

$$-\frac{|((x - y), h)|^2}{\|h\|^2} \geq 0, \quad \text{i.e.} \quad ((x - y), h) = 0. \qquad \square$$

Corollary B.2. *Let H be a Hilbert space, and let L be a Hilbert subspace.*

Then, $\forall x \in H$, $\exists y \in L$, $\exists z \perp L$ such that $x = y + z$. This presentation is unique.

Moreover, $\|x\|^2 = \|y\|^2 + \|z\|^2$ (Pythagoras theorem).

Proof.

1. Define $z = x - y$, where y is the element from Corollary 1.1, such that $\|x - y\| = \text{dist}(y, L)$ $(x \notin L)$.
   Theorem B.2 yields $x = y + z$, $z \perp L$.
   If $x \in L$, then $z = 0$ and $y = x$.
2. Let us show that this decomposition of x is unique.
   Assume that $x = y + z = y^* + z^*$, where $y, y^* \in L$ and $z, z^* \perp L$.
   So, $y - y^* = z^* - z$.
   But $y - y^* \in L$ and $z^* - z \in L^\perp$, so, $y - y^* = 0$ and $z^* - z = 0$ because $L \cap L^\perp = \{0\}$.
3. $\|x\|^2 = \|y+z\|^2 = ((y+z),(y+z)) = (y,y)+(y,z)+(z,y)+(z,z) = \|y\|^2 + \|z\|^2$ because $(y,z) = (z,y) = 0$ $(z \perp y)$. $\qquad\square$

Theorem B.3. *Let H be a Hilbert space, and let L be a subspace of H.*
  Then, $L^\perp$ is a Hilbert subspace of H.

Proof.

1. Let us prove that $L^\perp$ is a subspace of H.
   Indeed, $0 \in L^\perp$ because $0 \perp L$ $((0, x) = 0, \forall x \in L)$.
   Moreover, if $x, y \in L^\perp$, $\alpha, \beta \in \mathbb{K}$, then $\forall h \in L$, $(x, h) = 0$, $(y, h) = 0$, so $((\alpha x + \beta y), h) = 0$.
2. Let us prove that $L^\perp$ is complete.
   Let $(x_n)_{n \in \mathbb{N}} \subset L^\perp$ be a Cauchy sequence.
   Then, $\exists y \in H$ such that $x_n \to y$ (H is complete).
   Let us prove that $y \in L^\perp$.
   Indeed, $\forall n \in \mathbb{N}$, $\forall h \in L$, $(x_n, h) = 0$.
   So, $\lim_{n \to +\infty} (x_n, h) = (y, h) = 0 \Rightarrow y \in L^\perp$. $\qquad\square$

Proof of Theorem 1.6 for $\mathbb{K} = \mathbb{R}$.

1. Consider $\text{Ker} f = L = \{x \in H; f(x) = 0\}$. It is a Hilbert subspace.
   Indeed, it is a vector subspace of H.
   Let us check that L is closed.
   If $(x_n)_{n \in \mathbb{N}} \subset L$ converges to l, then by continuity of f, $(f(x_n))_{n \in \mathbb{N}}$ converges to $f(l)$.

As $x_n \in \text{Ker} f$, $f(x_n) = 0$, so $f(l) = 0$ and $l \in L$.

2. If $L = H$, then $f \equiv 0$, and we set $y = 0$.

   If $L \neq H$, then there exists $z_0' \in H \backslash L$.

   Let us construct z_0 such that $z_0 \perp L$, $z_0 \neq 0$, $f(z_0) = 1$.

   By Corollary B.1, $\exists y_0' \in L$ such that $\|z_0' - y_0'\| = \text{dist}(z_0', L)$ and $z_0' - y_0' \perp L$.

   By linearity, $f(z_0' - y_0') = f(z_0') - f(y_0') = f(z_0') \neq 0$ because $z_0' \notin L$.

   Define

   $$z_0 = \frac{z_0' - y_0'}{f(z_0')}.$$

   We have

   $z_0 \perp L$, $z_0 \neq 0$ because $z_0' \notin L$, and so $z_0' \neq y_0'$ and $f(z_0) = \dfrac{f(z_0') - f(y_0')}{f(z_0')} = 1$.

3. Let us prove that $\forall x \in H$, $f(x) = (x, y)$ with $y = \dfrac{z_0}{\|z_0\|^2}$.

   Indeed, if $x \in H$, then $x - f(x)z_0 \in L$ because $f(x - f(x)z_0) = f(x) - f(x)f(z_0) = 0$.

   Then, $x - f(x)z_0 \perp z_0$, so $(x, z_0) - f(x)\|z_0\|^2 = 0$,

   so, $f(x) = \dfrac{(x, z_0)}{\|z_0\|^2}$, i.e. $\forall x \in H$, $f(x) = (x, y)$ with $y = \dfrac{z_0}{\|z_0\|^2}$.

4. Let us prove the uniqueness of y.

   Indeed, assume that $\forall x \in H$, $f(x) = (x, y) = (x, y')$, and so $(x, (y - y')) = 0 \; \forall x \in H$.

   In particular, for $x = y - y'$, we have $\|y - y'\| = 0$, and so $y = y'$.

5. Let us prove that $\|f\| = \|y\|$.

   Indeed, by the Cauchy–Bunyakovsky–Schwarz inequality,

   $$|f(x)| = |(x, y)| \leq \|x\|\|y\|,$$

   so $\forall x \in H \backslash \{0\}$,

   $$\frac{|f(x)|}{\|x\|} \leq \|y\|, \quad \text{and so} \quad \|f\| \leq \|y\|.$$

Using $f(x) = (x, y)$, $\forall x \in H$, we take $x = y$ and get

$$|f(y)| = (y, y) = \|y\|^2, \quad \text{i.e.} \quad \exists x = y \quad \text{such that} \quad \frac{|f(y)|}{\|y\|} = \|y\|.$$

So, $\|y\|$ is the upper bound of $\left\{ \dfrac{|f(x)|}{\|x\|}, x \in H \backslash \{0\} \right\}$.

So, $\|y\| = \|f\|$. $\qquad\qquad\qquad\qquad\qquad\qquad\qquad\qquad\qquad$ $\square$

References

[1] L. Ait Mahiout, G. Panasenko, and V. Volpert. Homogenization of the diffusion equation with a singular potential for a model of a biological cell network, *Z. Angew. Math. Phys.*, 71: 181, 2020.

[2] L. Ait Mahiout, G. Panasenko, and V. Volpert. Partial homogenization of the diffusion equation with a Dirac-like potential. *J. for Multiscale Comp. Engineering*, 18: 5, 507–518, 2020.

[3] D. Cioranescu and F. Murat. Un terme étrange venu d'ailleurs, in *Nonlinear Partial Differential Equations and their Applications, Collège de France Seminar*, Vols. II and III, ed. by H. Brezis and J.-L. Lions, Research Notes in Mathematics, 60 and 70, Pitman, London, 1982, 93–138 and 154–178.

[4] V.A. Marchenko and E. Ya. Khruslov. *Boundary Value Problems in Domains with Fine-Grained Boundary.* Kiev: Naukova Dumka, 1974.

[5] Panasenko, G., The partial homogenization: Continuous and semidiscretized versions, *Mathematical Models and Methods in Applied Sciences*, 8: 17, 1183–1209, 2007.

[6] V.A. Trenogin. *The functional analysis*, Nauka, Moscow, 1980; French translation: V.A. Trenoguine. *Analyse fonctionnelle*, Éditions Mir, Moscou, 1985.

$$\text{Thus } \langle \Delta, \phi \rangle = \langle \Lambda, \phi \rangle \quad \forall \phi \text{ so take } \Delta = \Lambda \text{ for } \psi \text{ and for}$$

$$\langle \Delta \psi, \phi \rangle = \langle \psi, \phi \rangle \quad \langle \phi, \Sigma \rangle = \psi \quad \langle \Lambda, \phi \rangle = \frac{d\Delta \phi}{dM}$$

So, $\langle M, \phi \rangle$ is the upper bound of $\left[\int_{\Omega} \frac{d\Delta \phi}{dM} \right] + \nabla \Delta \cdot \nabla \phi \, dx$

$$\text{so } \langle M, \phi \rangle = \langle \Gamma, \phi \rangle$$

References

[1] Bear, Metabolic Considerations and Their Impact on Biological Studies, including computer simulations with transient potential intracellular and extracellular in nervous tissue, Neuro-regulation, 1997, 192-212.

[2] Le Rénbléfense, Ranancheva and V. ... ... : Partial homogenization of the diffusion equation with ... ... applied A. for electrode ... , Comm. Computing Res. J. Short, 1990.

[3] Clements and P. ... ... the nonnegative temperature distribution ... in two-dimensional finite element ... . Application ... ... energy sources. Vol. II and III. No. 1. ... First, no 1-6. Mono-graph Note ... , Mathematics, 60. ... ... , 1992, 89-98 and 102-15.

[4] J. L. Marchant and F. V. ... , Banach, ... , Bahn ... Elliptic Boundary ... Foundations. Boundless Kling Centric Paul in Bounded ... , ... , The fundamental solutions ... ... , ... , Advances in Mathematics of applications ... , for ... ... , 42 (4), ... nos. 3-4 (11), 100-1500, 2001.

[5] C. Marchant and ... the fundamental analytical ... in the Morgan ... , legal ... . to our ... for ... homogenization and the ... transaction, Dept. of ... at ... alb. , Mathematics, 1993.

Index

www.ingramcontent.com/pod-product-compliance
Lightning Source LLC
Chambersburg PA
CBHW050629190326
41458CB00008B/2203